CONTEMPORARY'S
REAL NUMBERS
Developing Thinking Skills in Math
Tables, Graphs, and Data Interpretation

Allan D. Suter

Project Editor
Kathy Osmus

CONTEMPORARY BOOKS

a division of NTC/CONTEMPORARY PUBLISHING GROUP
Lincolnwood, Illinois USA

ISBN: 0-8092-4217-6

Published by Contemporary Books,
a division of NTC/Contemporary Publishing Group, Inc.,
4255 West Touhy Avenue,
Lincolnwood (Chicago), Illinois, 60646-1975 U.S.A.
© 1991 by Allan D. Suter
8 9 0 DBH   18 17 16 15 14 13 12 11

Editorial Director
Caren Van Slyke

Editorial
Ellen Frechette
Sarah Conroy
Seija Suter
Karin Evans
Steve Miller
Karen Schenkenfelder
Ree Kline
Robin O'Connor
Laura Larson
Lisa Dillman

Editorial Production Manager
Norma Fioretti

Cover Design
Lois Koehler

Illustrator
Ophelia M. Chambliss-Jones

Art & Production
Marilyn Jusko

Typography
J•B Typesetting
St. Charles, Illinois

Cover photo by Michael Slaughter

CONTENTS

1 TABLES
Reading Tables 1
Calendars Are Tables 2
Working with a Schedule 3
Finding Bus Fares 4
Compare Rates Between Tables 5
Loan Payment Schedule 6
More Work with Loan Schedules 7
Comparing Prices 8
Using a Highway Mileage Chart 9
Postage-Rate Table 10
Sales Tax Table 11

2 PICTOGRAPHS
Comparing Information 12
Finding Information 13
Using Graphs to Make Decisions 14
Graph the Information 15
Using Tables and Graphs 16

3 CIRCLE GRAPHS
Working with Circle Graphs 17
Comparing Circle Graphs 18
From Table to Graph 19
Show the Percents 20
Earnings Statement 21

4 HORIZONTAL BAR GRAPHS
Reading Horizontal Bar Graphs 22
Comparing Values 23
Compare the Bars 24
Estimate the Percent 25
Graphing Monthly Expenses 26
Build a Horizontal Bar Graph 27

5 VERTICAL BAR GRAPHS
Reading Vertical Bar Graphs 28
Comparing Values 29
Compare the Bars 30
Graphing Monthly Savings 31
Build a Vertical Bar Graph 32

6 DOUBLE-BAR GRAPHS
Reading a Double-Bar Graph 33
Build a Double-Bar Graph 34

7   LINE GRAPHS

Reading a Line Graph 35
Find the Values 36
Plot the Points 37
Practice Helps 38
Showing General Trends 39

8   DOUBLE-LINE GRAPHS

Reading a Double-Line Graph 40
Build a Double-Line Graph 41
Graph the Data 42
Estimate the Distance 43

9   TALLY TABLES/FREQUENCY TABLES/HISTOGRAMS

Reading a Tally Table 44
Frequency Tables 45
Histogram 46
Working with a Histogram 47
Grouping Data Using a Histogram 48
Line Plots 49
Plot the Numbers 50

10   MEAN, OR AVERAGE

Learning about Averages 51
Find the Mean 52
Find the Average 53
Averaging Sales 54
Graph the Sales 55

11   RANK, MEDIAN, RANGE, AND MODE

Rank Order 56
Median and Mode 57
Find the Median 58
Plot the Median 59
Mean, Median, and Range 60

12   APPLYING TABLES, DATA, AND GRAPHS

Using Data 61
Finding Data 62
Graph the Data 63
Comparing Graphs 64
Practice Your Skills 65

GRAPHS REVIEW 66

ANSWER KEY 67

Reading Tables

Tables help you find information you need quickly and easily. They organize information using words and numbers.

Title →
What is the table about?

Buildings in Toronto	
Building	**Number of Stories**
A	40
B	30
C	5
D	15

← **Headings**
What information is in the table?

↑ **Column** ↑
What information is given from top to bottom, under each heading?

▶ Use the **title, headings,** and **columns** to find what information a table contains.

1. How many buildings are there? _____

2. How many stories are in the tallest building? _____

To compare information from column to column, read the **rows.**

Buildings in Toronto	
Building	**Number of Stories**
A	40
B	30
C	5
D	15

Row
What information is given from left to right, under each heading?

← Read across:
Building B, 30 stories

3. How many stories does Building B have? _____

4. List the buildings that are over 10 stories tall. _____

5. What building is 5 stories tall? _____

Calendars Are Tables

Calendars are tables that we use every day.

July						
Sun.	Mon.	Tue.	Wed.	Thu.	Fri.	Sat.
	1	2	3	4	5	6
7	8	9	10	11	12	13
14	15	16	17	18	19	20
21	22	23	24	25	26	27
28	29	30	31			

▶ Use the calendar above to answer questions 1–4.

1. July 4 falls on what day of the week? _____

2. What is the date of the second Saturday in the month? _____

3. Two weeks from July 12 is what date? _____

4. How many Wednesdays are there in July? _____

August						
Sun.	Mon.	Tue.	Wed.	Thu.	Fri.	Sat.

▶ Use the August calendar to answer questions 5–8.

5. Fill in the numbers for the month of August. Assume that August 1 is on a Thursday. August has 31 days.

6. August 28 falls on what day of the week? _____

7. The fourth Friday of the month is what date? _____

8. Three weeks from August 6 is what date? _____

Working with a Schedule

Airline flights are given flight numbers to make it easier to keep track of schedules.

Title ⟶ **From Chicago (O'Hare) to Atlanta**

Flight #	Departure	Arrival
133	7:35a	10:31a
886	9:58a	12:54p
180	1:15p	4:06p
132	3:55p	6:51p
525	8:20p	11:18p

Remember to read the headings.

a   stands for A.M., which is the time from midnight to noon.

p   stands for P.M., which is the time from noon to midnight.

Find 180. ⟶

↑
Find where the departure column and the row with 180 come together.

Example

What is the departure time for Flight #180?

<u>Step 1</u>

Find the column that lists flight numbers and find #180.

<u>Step 2</u>

Then follow the row that holds Flight #180 across to the departure column.

The departure time for Flight #180 is 1:15 P.M.

▶ Use the airline schedule to answer the following questions. Remember that Atlanta is on Eastern Standard Time and Chicago is on Central Standard Time.

1. How many flights are there to Atlanta? _____

2. What is the earliest time you could leave (depart) Chicago? _____

3. What is the latest time you could arrive in Atlanta? _____

4. What is the departure time for Flight #886? _____

5. What is the arrival time for Flight #132? _____

6. What is the flying time (the time between arrival and departure) for Flight #525? _____

Finding Bus Fares

Bus Fares from Twin Falls			
To	**One-Way Ticket**	**Round-Trip Ticket**	**Five-Trip Ticket***
Jerome	$3.35	$5.90	$13.25
Gooding	4.75	7.45	20.45
Rupert	5.65	9.95	24.30
Eden	2.50	4.50	10.70

Find Rupert. ⟶ (Rupert row)

*Five-trip ticket = 5 one-way tickets.

↑ Find the round-trip ticket column.

Example

How much does a round-trip ticket to Rupert cost?

A round-trip ticket to Rupert costs $9.95.

Step 1

Find the column that lists where the buses are going. Find Rupert.

Step 2

Find the place where the round-trip ticket column meets the row that Rupert is in.

▶ Use the information from the table to answer the following questions.

1. How much more is a one-way ticket to Gooding than a one-way ticket to Eden? _____

2. How much more is a five-trip ticket to Rupert than a round-trip ticket to Jerome? _____

3. a) How much more is a round-trip ticket to Jerome than a one-way ticket? _____

   b) How much do 2 one-way tickets to Jerome cost? _____

   c) Which costs less: a round-trip ticket or 2 one-way tickets to Jerome? _____

4. a) How much would 5 round-trip tickets to Rupert cost? _____

   b) How much would 2 five-trip tickets to Rupert cost? _____

   c) Which costs less: 5 round-trip tickets or 2 five-trip tickets to Rupert? _____

Compare Rates Between Tables

Rates for Ads in the Weekly Shopper	
Size	**Rates**
12-inch ad	$45.00
8-inch ad	30.00
6-inch ad	22.50
4-inch ad	15.00
2-inch ad	7.50

Rates for Ads in the Daily Newspaper	
Size	**Rates**
12-inch ad	$33.00
8-inch ad	22.00
6-inch ad	16.50
4-inch ad	11.00
2-inch ad	5.50

←— compare rates —→

Example

Which costs more: a 12-inch ad in the weekly shopper or a 12-inch ad in the daily newspaper?

Step 1

Find the rate for a 12-inch ad in the weekly shopper ($45).

Step 2

Find the rate for a 12-inch ad in the daily newspaper ($33) and compare the rates.

$45 is more than $33, so a 12-inch ad in the weekly shopper costs more.

___

▶ Use the tables above to answer the following questions.

1. a) What is the cost of a 4-inch ad in the daily newspaper? _____

b) Are there any ads that cost less than that in the weekly shopper? _____

2. a) How much does an 8-inch ad in the daily newspaper cost? _____

b) What size ad in the weekly shopper costs about the same? _____

3. a) What size ad could you place in the daily newspaper for $10.00 or less? _____

b) Could you place the same size ad in the weekly shopper for under $10.00? _____

4. a) What size ads in the daily newspaper cost more than $20.00? _____

b) What size ads in the weekly shopper cost more than $20.00? _____

Loan Payment Schedule

When a bank loans money, the amount borrowed and the length of time needed to pay back the loan affect the total amount that must be repaid.

Example: How much would you have to pay each month if you borrowed $500 for 2 years? $23.78

Monthly Loan Payments					
Loan	1 Year	2 Years	3 Years	4 Years	5 Years
$ 25	$ 2.24	$ 1.19	$ .85	$ .68	$ .57
50	4.47	2.38	1.69	1.35	1.14
75	6.70	3.57	2.53	2.02	1.71
100	8.94	4.76	3.37	2.69	2.28
200	17.87	9.51	6.74	5.37	4.56
300	26.80	14.27	10.11	8.05	6.83
400	35.73	19.02	13.48	10.74	9.11
500	44.66	23.78	16.85	13.42	11.38
600	53.60	28.53	20.22	16.10	13.66
700	62.53	33.28	23.59	18.78	15.93
800	71.46	38.04	26.96	21.47	18.21
900	80.39	42.79	30.33	24.15	20.48
1,000	89.32	47.55	33.70	26.83	22.76
2,000	178.64	95.09	67.39	53.66	45.51
3,000	267.96	142.63	101.09	80.49	68.26

Find $500 in the loan column.

Find where the $500 row and the 2-year column meet.

▶ Use the table above to answer the following questions.

1. How much would you have to pay each month if you borrowed:

   a) $700 for 5 years? _____   **d)** $300 for 2 years? _____

   b) $3,000 for 3 years? _____   **e)** $900 for 4 years? _____

   c) $100 for 1 year? _____   **f)** $2,000 for 3 years? _____

2. a) How many monthly payments would you make in 1 year? _____

   b) How many monthly payments would you make in 2 years? _____

3. a) If you borrow $200 for 2 years, how much will you pay each month? _____

   b) What is the total amount you will pay in monthly payments? _____

4. a) You need to borrow $3,000 to buy a car. If you chose a 2-year loan, what is the total amount you would have to pay back? _____

   b) If you borrowed $3,000 and chose a 3-year loan, what is the total amount you would have to pay back? _____

More Work with Loan Schedules

Sometimes the amount you borrow and how long you take the loan out for depends on how much you can afford to pay each month.

Example

Jean wants to borrow $1,000. She can afford to pay about $50 per month. What is the shortest amount of time she could take a loan out for and still pay around $50 per month?

Loan	1 Year	2 Years
$ 25	$ 2.24	$ 1.19
50	4.47	2.38
75	6.70	3.57
100	8.94	4.76
200	17.87	9.51
300	26.80	14.27
400	35.73	19.02
500	44.66	23.78
600	53.60	28.53
700	62.53	33.28
800	71.46	38.04
900	80.39	42.79
1,000	89.32	47.55

loan for $1,000 →

less than $50 ↑

Jean could take out a 2-year loan and still afford the payments.

▶ Use the loan payment schedule on page 6 to help you answer the following questions.

1. Tim needs to borrow $500. He can afford to pay between $15 and $20 a month. What is the least amount of time he can take a $500 loan out for? _____

2. How many years would you take a $100 loan out for if you wanted monthly payments around $10? _____

3. How many years would you take a $700 loan out for if you wanted monthly payments around $15? _____

4. Mario wants to take out a 1-year loan, but he can't afford to pay more than $75 a month. What is the largest loan that he can afford? _____

5. If you wanted to take out a 3-year loan with monthly payments around $65, what is the largest loan that you could choose? _____

6. If you wanted to take out a 2-year loan with monthly payments around $90, what is the largest loan that you could choose? _____

Comparing Prices

A table can show a wide variety of information that will help you compare items quickly and find out what you want to know.

Survey of American and Japanese Prices (in dollars)			
Product	**Product Origin**	**Price in Tokyo**	**Price in New York**
Blue jeans	United States	$55.63	$32.00
Movie ticket	Japan, United States	11.27	7.25
Pizza	Japan, United States	14.08	8.00
Calculator	Japan	4.79	7.00
Tires (one)	Europe	72.18	87.00
Bed linens	Australia	63.38	20.00

Source: Basic data from U.S. Department of Commerce, Japanese Ministry of International Trade and Industry.

▶ Use the table above to answer the following questions.

1. What is the table about? _____

2. List where each product is made.

   a) Bed linens _____   **d)** Movie ticket _____

   b) Pizza _____   **e)** Tire _____

   c) Blue jeans _____   **f)** Calculator _____

3. What products cost more in Tokyo than in New York? _____
   _____

4. What products cost more in New York than in Tokyo? _____
   _____

5. How much more does it cost to buy a movie ticket in Tokyo than in New York? _____

Using a Highway Mileage Chart

The mileage chart below shows how far it is in miles from one city to another.

Mileage Chart*

	Seattle	Miami	Chicago	Detroit	Las Vegas	New York City	Tampa	Houston	Boston
Seattle		3,303	2,052	2,327	1,180	2,841	3,077	2,369	3,016
Miami	3,303		1,397	1,385	2,570	1,334	250	1,190	1,520
Chicago	2,052	1,397		275	1,780	809	1,160	1,091	994
Detroit	2,327	1,385	275		2,020	649	1,184	1,276	799
Las Vegas	1,180	2,570	1,780	2,020		2,572	2,299	1,467	2,752
New York City	2,841	1,334	809	649	2,572		1,150	1,610	208
Tampa	3,077	250	1,160	1,184	2,299	1,150		943	1,361
Houston	2,369	1,190	1,091	1,276	1,467	1,610	943		1,830
Boston	3,016	1,520	994	799	2,752	208	1,361	1,830	
Atlanta	2,625	663	708	732	1,979	854	452	791	1,108

*Distances are shown in miles.

Example: How far is it from Atlanta to Boston?

Step 1
Find the row for Atlanta.

Step 2
Find the column for Boston.

Step 3
Follow along the row and column until they meet.

Atlanta is 1,108 miles from Boston.

▶ Complete the questions using the mileage chart above.

1. What is the mileage from New York City to Chicago? _____

2. What is the mileage from Houston to Las Vegas? _____

3. Find the mileage for a round-trip between Detroit and Houston. _____

4. How much farther is it from Miami to Las Vegas than from Houston to Tampa? _____

5. What is the distance of a trip from Detroit to Chicago to Miami? _____

Postage-Rate Table

Postal Rates							
Weight Not to Exceed	Postage-Rate Zones						
	2	3	4	5	6	7	8
1 lb.	$1.47	$1.59	$1.75	$1.83	$1.91	$2.01	$2.10
2	1.48	1.61	1.96	2.08	2.25	2.41	2.60
3	1.56	1.75	2.13	2.31	2.55	2.80	3.05
4	1.64	1.89	2.31	2.49	2.81	3.12	3.46
5	1.73	1.98	2.40	2.63	3.01	3.39	3.79
6	1.82	2.05	2.49	2.74	3.14	3.58	4.04
7	1.90	2.12	2.59	2.86	3.32	3.82	4.32
8	1.99	2.18	2.70	3.08	3.61	4.16	4.74
9	2.09	2.31	2.87	3.29	3.90	4.51	5.17
10	2.17	2.43	3.03	3.50	4.17	4.87	5.58

When mailing a package, you are charged according to zones. Zone 5 is farther away than Zone 3, so you must pay more.

"Weight not to exceed" means that if a package weighs more than the given number of pounds, you must round its weight up to the next highest pound.

Example A: 1 pound 8 ounces must be rounded to 2 pounds.
Example B: 7 pounds 1 ounce must be rounded to 8 pounds.

▶ Use the postage-rate table to find the following shipping costs.

	Postage Weight	Zone	Shipping Cost
1.	6 pounds	5	
2.	2 pounds	7	
3.	1 pound 3 ounces	3	
4.	9 pounds 8 ounces	2	
5.	2 pounds 11 ounces	8	
6.	5 pounds 15 ounces	6	

7.

Zone 4
2 lb. 8 oz.

Zone 7
5 lb. 3 oz.

Find the total shipping cost of the 2 packages. _____

8.

Zone 8
8 lb. 7 oz.

Zone 5
4 lb. 5 oz.

Find the total shipping cost of the 2 packages. _____

Sales Tax Table

Work gloves
$2.85

Duct tape
$2.35

Masking tape
$.59

Paintbrush
$3.48

Flood lamp
$1.88

Sponge
$.99

Sales Tax Table	
Amount	Tax
$3.92 to $4.08	$.24
4.09 to 4.24	$.25
4.25 to 4.41	$.26
4.42 to 4.58	$.27
4.59 to 4.74	$.28
4.75 to 4.91	$.29
4.92 to 5.08	$.30
5.09 to 5.24	$.31
5.25 to 5.41	$.32
5.42 to 5.58	$.33
5.59 to 5.74	$.34
5.75 to 5.91	$.35

▶ Follow these steps to complete each problem:
- Find the price for each item and add to find the **subtotal.**

- Find the amount of the subtotal on the sales tax table and locate the **sales tax** for that amount.

- Add the subtotal and sales tax to find the **total** price.

1.

Value Hardware		
Item	Price	
1 Paintbrush		
1 Flood lamp		
Thank You	Subtotal	
	Tax	
	Total	

3.

Value Hardware		
Item	Price	
2 Masking tapes		
1 Pair of work gloves		
Thank You	Subtotal	
	Tax	
	Total	

2.

Value Hardware		
Item	Price	
3 Sponges		
1 Duct tape		
1 Masking tape		
Thank You	Subtotal	
	Tax	
	Total	

4.

Value Hardware		
Item	Price	
2 Duct tapes		
Thank You	Subtotal	
	Tax	
	Total	

Comparing Information

A **pictograph** shows comparisons of numbers using symbols, or pictures. Each symbol stands for a certain number of things, often using rounded numbers.

Guidelines for Understanding Pictographs
- Read the **title** to know what the graph is about.
- Look at the **key,** which tells what each symbol means.
- Study the comparisons.

Favorite Forms of Exercise	
Exercise	**Number of People**
Walking	👤 👤 👤 👤 👤 👤 👤 👤
Jogging	👤 👤 👤
Aerobics	👤 👤 👤 👤 👤
Exercising with equipment	👤 👤　　Key: 👤 = 5 million people

Title → (points to Favorite Forms of Exercise)
Column Heading ← (points to Number of People)
Key ← (points to Key)

Example

How many people chose jogging as their favorite form of exercise?

Step 1

Find jogging in the exercise column.

Step 2

Count how many symbols are in that row.

Step 3

Since 1 symbol equals 5 million, 3 symbols equal 15 million.

15 million people chose jogging as their favorite form of exercise.

▶ Use the pictograph to answer the following questions.

1. What is the title of the pictograph? _____

2. How many people does each symbol represent? _____

3. How many forms of exercise are listed? _____

4. What is the least popular form of exercise? _____

5. What is the most popular form of exercise? _____

6. What is the second most popular form of exercise? _____

7. How many more people chose walking over jogging? _____

Finding Information

Pictographs are useful when comparing general information. If more specific amounts are needed, they can be calculated using the key to the graph.

Cost of a Pound of Apples in Selected World Capitals	
World Capitals	**Cost per Pound**
Washington, D.C.	🍎 🍎 🍎 🍎 ← $\frac{1}{2}$ of an apple equals $.15 per pound
London	🍎 🍎 🍎
Pretoria	🍎 Key: 🍎 = $.30 per pound
Stockholm	🍎 🍎 🍎

Source: Department of Agriculture.

Example: How much do apples cost per pound in London?

Step 1
Find London under the heading of world capitals.

Step 2
Count how many symbols are listed for London.

$2\frac{1}{2}$

Step 3
Check the key for the value of each symbol and find the amount shown for London.

$.30 + $.30 + $.15 = $.75

Apples cost $.75 per pound in London.

▶ Use the pictograph to answer the following questions.

1. What is the title of the pictograph? _____

2. How many world capitals are listed? _____

3. How much money per pound does each symbol 🍎 represent? _____

4. Sometimes half of a symbol is used in a pictograph. If 🍎 = $.30,

   then ◖ = _____ .

5. In what world capital are apples the most expensive?
   (Hint: Look for the most symbols.) _____

6. In what world capital are apples the least expensive? _____

7. How much do apples cost per pound in Pretoria? _____

Using Graphs to Make Decisions

Sometimes graphs give us information that we can use to make decisions. For example, the graph below shows the average waiting time to have a car's exhaust output checked. The waiting time might help you decide what day to take your car in.

Average Waiting Time for Emissions Check

Monday	🚗🚗🚗
Tuesday	🚗
Wednesday	🚗🚗
Thursday	🚗
Friday	🚗🚗🚗
Saturday	🚗🚗🚗🚗

Key: 🚗 = 10 minutes

Example: What day has an average waiting time of 10 minutes?

Step 1	Step 2	Step 3
Find out how much time each symbol (car) represents.	Find the row with only 1 car in it.	Find the day that goes with that row.
10 minutes	1 car = 10 minutes	Tuesday

Tuesday has an average waiting time of 10 minutes.

▶ Use the pictograph to answer the following questions.

1. How long is the wait on Wednesday? _____

2. How long is the wait on Monday? _____

3. What day has the shortest waiting time? _____

4. How much longer is the wait on Saturday than on Friday? (Hint: Just count how many more symbols there are on Saturday and find their value.) _____

5. a) Which day has the longer wait, Tuesday or Wednesday? _____

    b) How much longer is the wait? _____

6. What day has an average waiting time of 30 minutes? _____

Graph the Information

Sometimes information from a table can be shown on a graph.

Average Time Spent at Events	
Attraction	**Average Length of Activity**
Amusement park	5 hours
Ice hockey game	3.5 hours
Horse races	4 hours

◄————— .5 equals $\frac{1}{2}$ an hour

▶ Use the information from the table above to help you complete the pictograph.

1.

Average Time Spent at Events	
Attraction	**Average Length of Activity**
NFL football game	🕐 🕐 🕐
Amusement park	
Ice hockey game	
Horse races	🕐 🕐 🕐 🕐

Key: 🕐 = 1 hour

Source: International Food Service Manufacturers' Association.

▶ Use the pictograph to answer the following questions.

2. What is the pictograph about? _____

3. a) How did you show .5 of an hour? _____

   b) How many minutes is .5 of an hour? _____

4. On the average, how much more time is spent visiting an amusement park than watching a football game? _____

5. On the average, how much more time is spent watching horse races than a hockey game? _____

Using Tables and Graphs

▶ Use the information from the table to complete the pictograph. Be sure to check how many people each symbol represents.

Largest Cities in the World by the Year 2000	
City	Population (in millions)
New York, USA	15
Tokyo, Japan	30
Seoul, South Korea	20
Mexico City, Mexico	25

1.

Largest Cities in the World by the Year 2000	
City	Population (in millions)
New York, USA	🧍 🧍 🧍
Tokyo, Japan	
Seoul, South Korea	
Mexico City, Mexico	
Key: 🧍 = 5,000,000 (5 million) people	

Basic Data: U.S. Census Bureau.

▶ Use the pictograph to answer the following questions.

2. Each symbol 🧍 represents how many people? _____

3. What city will be twice the size of New York in 2000? _____

4. Of the 4 cities shown, which city will probably have the smallest population? _____

5. By the year 2000, what will be the 2 largest cities in the world? _____

Working with Circle Graphs

A **circle graph** is a graph that shows the whole amount as a circle. Sometimes they are called pie charts because of their shape.

Circle graphs
- represent a whole divided into parts
- make it easy to compare parts
- most often show percents, fractions, or decimals

▶ Answer the following questions based on the graphs.

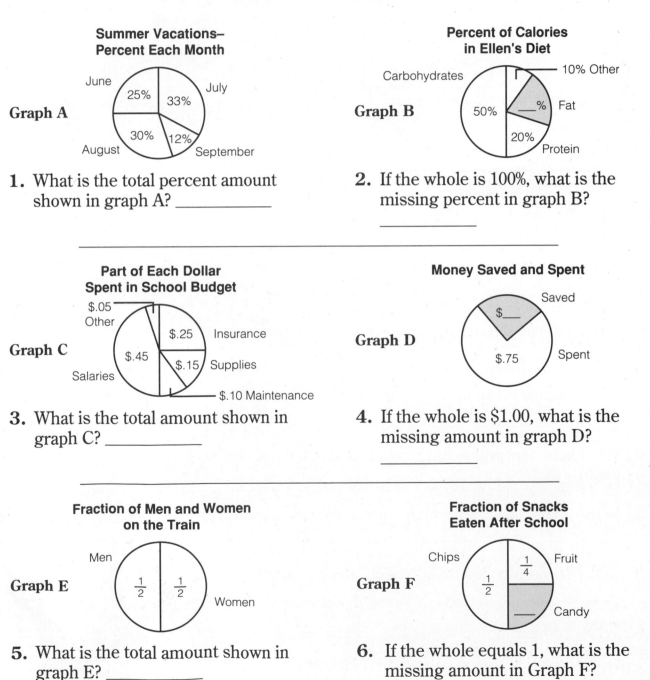

1. What is the total percent amount shown in graph A? _____

2. If the whole is 100%, what is the missing percent in graph B?

_____

3. What is the total amount shown in graph C? _____

4. If the whole is $1.00, what is the missing amount in graph D?

_____

5. What is the total amount shown in graph E? _____

6. If the whole equals 1, what is the missing amount in Graph F?

_____

Comparing Circle Graphs

Circle graphs can be used to compare information.

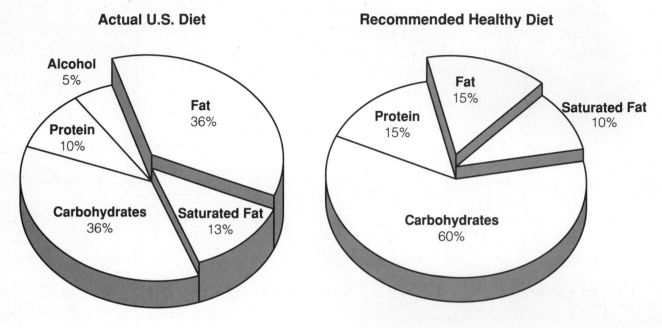

Actual U.S. Diet

Alcohol 5%
Protein 10%
Fat 36%
Carbohydrates 36%
Saturated Fat 13%

Recommended Healthy Diet

Fat 15%
Protein 15%
Saturated Fat 10%
Carbohydrates 60%

Source: National Academy of Sciences.

▶ Use the circle graphs to answer the following questions.

1. What 3 items make up the largest portion of the actual U.S. diet? _____
_____

2. What 3 items make up the largest portion of the recommended healthy diet? _____

3. What item appears on the actual U.S. diet but not on the recommended healthy diet? _____

4. In the actual U.S. diet, which 2 items are equal in size? _____
_____

5. In the recommended healthy diet, which 2 items are equal in size? _____
_____

6. When comparing the actual U.S. diet to the recommended healthy diet, what 2 items show the greatest difference? _____

From Table to Graph

Information from tables can be shown in the form of circle graphs.

Mike's Monthly Earnings		
Job	**Money Earned**	**Percent of Earnings**
Lawn mowing	$ 36.00	30%
Delivering papers	24.00	20
Fix-it jobs	12.00	10
Painting	48.00	40
Total	120.00	100

1. On the circle graph, fill in the percent for each job category.

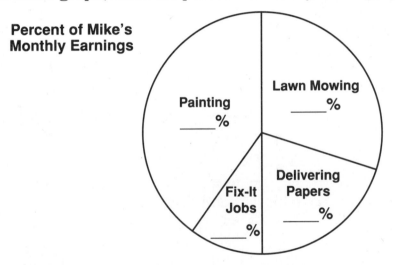

Percent of Mike's Monthly Earnings

▶ Answer the questions using the monthly earnings table and circle graph.

2. Which job earned the most money? _____

3. Fix-it jobs brought in half the earnings of what other job category? _____

4. Together, lawn mowing and painting made up what percent of Mike's earnings? _____

5. How much more did Mike earn by painting than by delivering papers? _____

6. State 2 facts about the circle graph. _____

_____

_____

Show the Percents

Circle graphs make it easy to compare the sizes of the parts that make up a whole.

▶ Use the information from the table to complete the circle graph.

Family Budget	
Item	**Percent**
Food and shelter	50%
Clothing	10
Savings	15
Other	25

1.

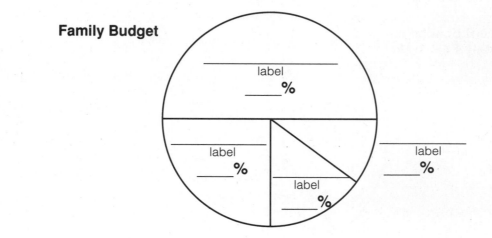

Family Budget

▶ Answer each question about the circle graph.

2. The whole circle represents 100%. What fraction of the whole circle does 25% represent? _____

3. What percent of the whole circle is represented by food and shelter and clothing combined? _____

4. The whole circle is divided into how many parts? _____

5. If you add the percentages for each of the 4 categories, what percent do you get? _____

6. a) What 2 categories add up to 25%? _____

b) Do they take up one-fourth of the whole circle? _____

7. State 2 facts about the circle graph. _____

_____

Earnings Statement

Below is Gail Smith's payment stub that shows how much money she earned **(gross pay)**, how much money was taken out **(deductions)**, and how much she took home **(net pay)**.

Name	Week Ending	Gross Pay	Net Pay
Gail Smith	5/3/90	$1,136.82	$704.83

Tax Deductions			Optional Deductions		
Federal	FICA	State	Medical	Union Dues	Others
$170.52	$90.95	$56.84			$113.68

The circle graph shows this information in another form. It shows the different parts of her payment stub in percent form. Remember that you can also read this graph as the percent of each dollar earned. So 8% of each dollar equals $.08.

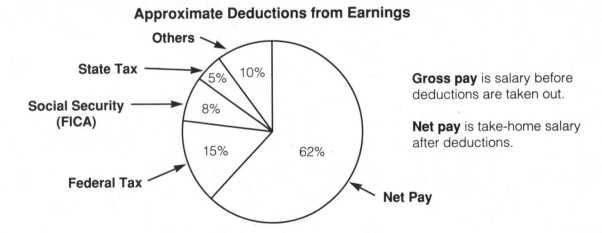

Approximate Deductions from Earnings

Others — 10%
State Tax — 5%
Social Security (FICA) — 8%
Federal Tax — 15%
Net Pay — 62%

Gross pay is salary before deductions are taken out.

Net pay is take-home salary after deductions.

▶ Use the circle graph to answer the following questions.

1. For every dollar Gail earns:
   a) __62__ cents is net pay
   b) _____ cents is federal tax
   c) _____ cents is social security (FICA)
   d) _____ cents is state tax
   e) _____ cents is other deductions

2. Do the net pay and all of the deductions add up to $1.00 or 100% of the entire circle? _____

3. For every $100 she earns, Gail will take home (net pay) how many dollars after deductions? _____

Reading Horizontal Bar Graphs

A common way to represent numbers or amounts is with bars. Bars that run from left to right are called **horizontal bars.** (They are level like the horizon.) Lines that mark the length of the bars are called **horizontal scales.**

▶ Use the horizontal bars to answer the following questions.

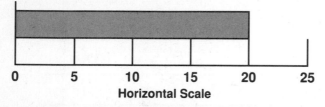

Horizontal Scale

1. What number does the horizontal bar represent? _____20_____

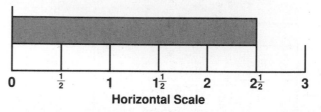

Horizontal Scale

3. What number does the horizontal bar represent? _____

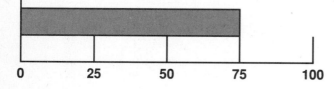

2. What number does the horizontal bar represent? _____

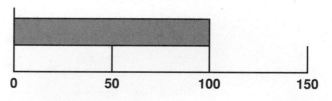

4. What number does the horizontal bar represent? _____

▶ Sometimes the bar stops at a place that is not labeled. In such cases, use the scale to help you judge the amount shown on the graph.

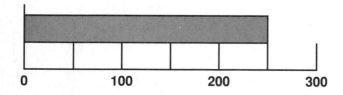

5. What number does the horizontal bar represent? _____250_____

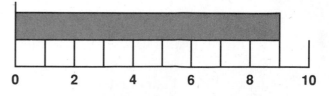

7. What number does the horizontal bar represent? _____

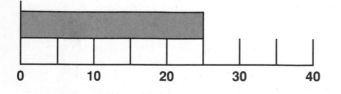

6. What number does the horizontal bar represent? _____

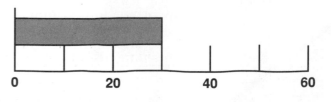

8. What number does the horizontal bar represent? _____

Comparing Values

A bar graph uses different lengths of bars to compare values. Horizontal bar graphs use bars that run across the page (horizontally).

▶ Use the graphs to answer the following questions.

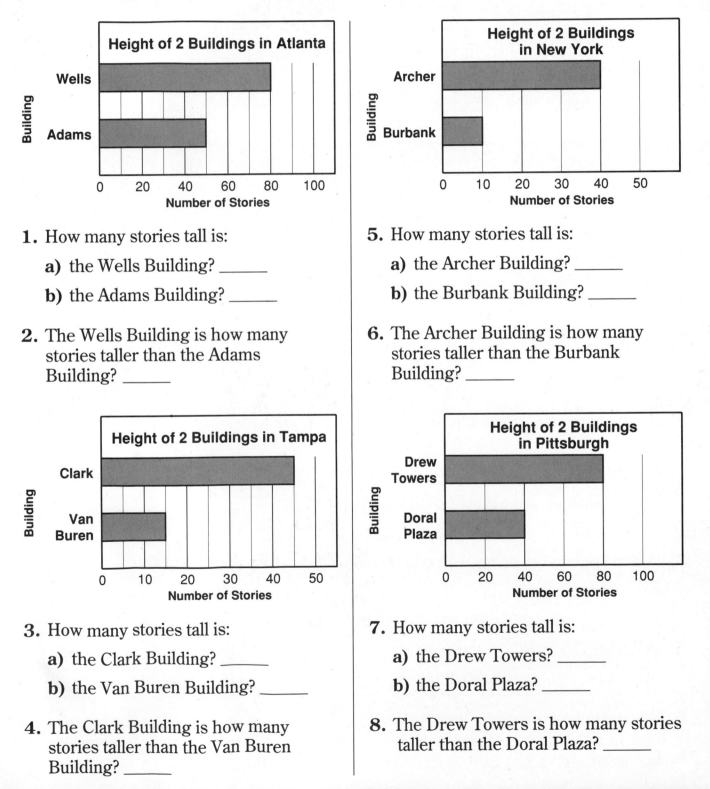

1. How many stories tall is:

a) the Wells Building? _____

b) the Adams Building? _____

2. The Wells Building is how many stories taller than the Adams Building? _____

3. How many stories tall is:

a) the Clark Building? _____

b) the Van Buren Building? _____

4. The Clark Building is how many stories taller than the Van Buren Building? _____

5. How many stories tall is:

a) the Archer Building? _____

b) the Burbank Building? _____

6. The Archer Building is how many stories taller than the Burbank Building? _____

7. How many stories tall is:

a) the Drew Towers? _____

b) the Doral Plaza? _____

8. The Drew Towers is how many stories taller than the Doral Plaza? _____

Compare the Bars

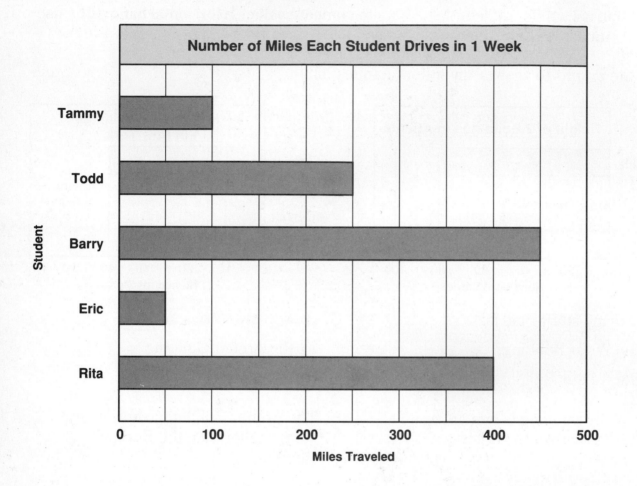

Number of Miles Each Student Drives in 1 Week

Student

Tammy

Todd

Barry

Eric

Rita

0   100   200   300   400   500

Miles Traveled

▶ Use the horizontal bar graph to answer the following questions.

1. What is the graph about? _____

2. Which students drive more than 200 miles each week? _____

_____

3. Which student drives more than 200 miles but less than 300 miles each
week? _____

4. How many more miles did Barry drive than:

    a) Tammy? _____   **c)** Eric? _____

    b) Todd? _____   **d)** Rita? _____

5. Which student drives the fewest miles? _____

6. In all, how many miles do all 5 students drive in 1 week? _____

Estimate the Percent

Sometimes a bar stops between 2 markings. In such cases, you need to judge the approximate value. Ask yourself questions such as, "Which labeled number is it closer to? Does it fall halfway between 2 labeled values?"

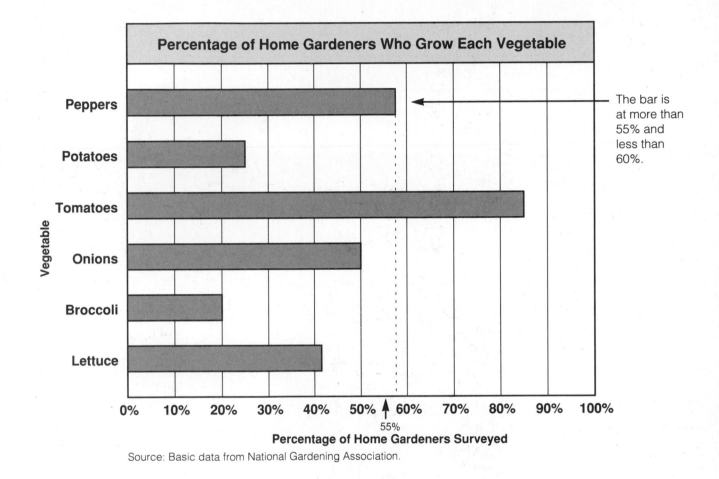

Source: Basic data from National Gardening Association.

▶ Use the horizontal bar graph to answer the following questions.

1. What is the bar graph about? _____

2. About what percent of home gardeners grow:

    a) Peppers? <u>58%</u>   **d)** Onions? _____

    b) Potatoes? _____   **e)** Broccoli? _____

    c) Tomatoes? _____   **f)** Lettuce? _____

3. Do more home gardeners grow broccoli or potatoes? _____

4. What are the 3 most popular vegetables grown by home gardeners? _____
_____

Graphing Monthly Expenses

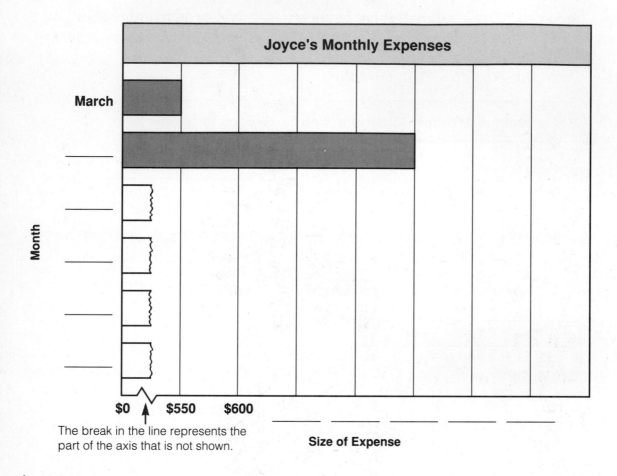

Joyce's Monthly Expenses

March

Month

$0 $550 $600

The break in the line represents the
part of the axis that is not shown.

Size of Expense

▶ Use the graph to answer the following questions.

1. Finish labeling the horizontal scale by writing the missing dollar amounts.

2. Finish labeling the vertical scale by writing the months April, May, June, July, and August.

3. Complete the bars to show that Joyce's expenses were $800 in May, $600 in June, $700 in July, and $650 in August.

4. In which month were Joyce's expenses the highest? _____

5. In which month were Joyce's expenses the lowest? _____

6. What is the difference in expenses between March and April? _____

7. How much more were the expenses in July than in August? _____

8. What were Joyce's total expenses for the 6 months? _____

Build a Horizontal Bar Graph

A bar graph is a way of visually comparing numbers. Follow the instructions below, using information from the table to build a horizontal bar graph.

1. Write the title of the graph above the graph.

2. Label the vertical scale with the days listed in the table.
   Try to make them equally spaced.

3. Label the horizontal scale by fifties (50, 100, 150 . . .) up to 500,
   with equal space between numbers.

4. Using the information from the table, draw a bar to show the number of people for each day. The bars should be the same width with equal space between the bars.

Shoppers at the Mall	
Day	**Number**
Monday	160
Tuesday	230
Wednesday	250
Thursday	400
Friday	450

Reading Vertical Bar Graphs

Bars often represent numbers or amounts. Bars that run up and down are called vertical bars. Lines that mark the length of the bars are called vertical scales.

▶ Use the vertical bars to answer the following questions.

▶ Remember that some bars may stop on lines that are not labeled. Use the numbers that are given to find the missing values.

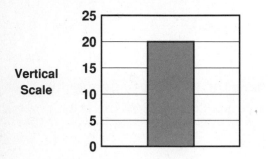

1. What number is represented by the vertical bar? __20__

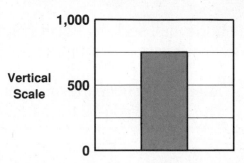

4. What number is represented by the vertical bar? _____

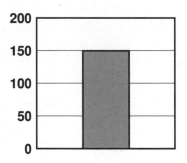

2. What number is represented by the vertical bar? _____

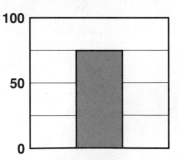

5. What number is represented by the vertical bar? _____

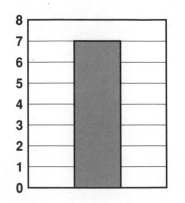

3. What number is represented by the vertical bar? _____

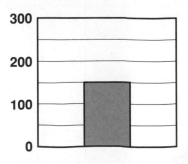

6. What number is represented by the vertical bar? _____

Comparing Values

Like a horizontal bar graph, a vertical bar graph uses different lengths of bars to compare values. Vertical bar graphs use bars that run up and down (vertically).

▶ Use the graphs to answer the following questions.

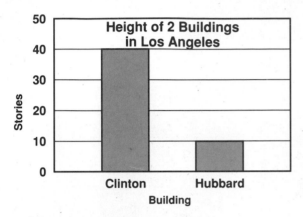

1. How many stories tall is:

  a) the Clinton Building? _____

  b) the Hubbard Building? _____

2. The Clinton Building is how many stories taller than the Hubbard Building? _____

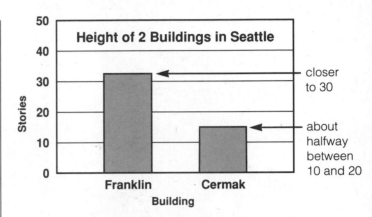

5. About how many stories tall is:

  a) the Franklin Building? _____

  b) the Cermak Building? _____

6. The Franklin Building is about how many stories taller than the Cermak Building? _____

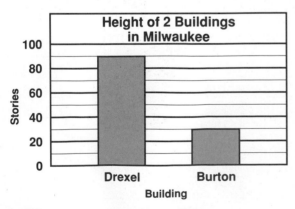

3. How many stories tall is:

  a) the Drexel Building? _____

  b) the Burton Building? _____

4. The Drexel Building is how many stories taller than the Burton Building? _____

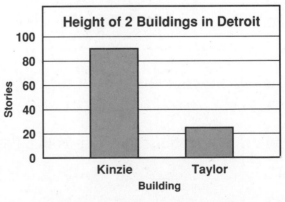

7. About how many stories tall is:

  a) the Kinzie Building? _____

  b) the Taylor Building? _____

8. The Kinzie Building is about how many stories taller than the Taylor Building? _____

Compare the Bars

Heights of U.S. Mountains

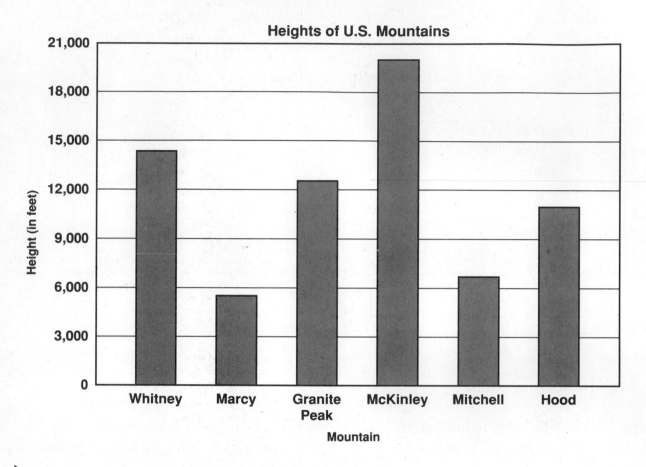

▶ Use the vertical bar graph to answer the following questions.

1. Name the mountains taller than 9,000 feet. _____
_____

2. Name the mountains shorter than 12,000 feet. _____
_____

3. Rank the mountains in order from tallest to shortest:

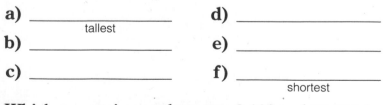

a) _____ **d)** _____
     tallest

b) _____ **e)** _____

c) _____ **f)** _____
                       shortest

4. Which mountains are between 3,000 and 12,000 feet tall? _____
_____

5. An airplane flies at 25,000 feet. Approximately what is the difference between the height of the airplane and the top of the tallest mountain? _____

Graphing Monthly Savings

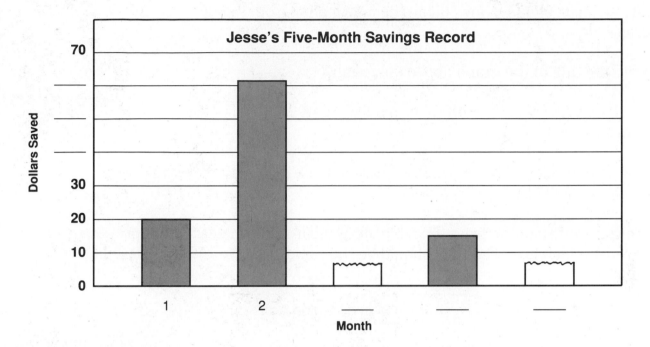

Jesse's Five-Month Savings Record

Dollars Saved

70

30

20

10

0

1   2   ___   ___   ___

Month

▶ Use the graph to answer the following questions.

1. Finish labeling the horizontal scale by writing the numbers to show the third, fourth, and fifth months.

2. Finish labeling the vertical scale showing dollar amounts in tens (10, 20, 30, . . .).

3. Make bars to show that Jesse saved $42 in the third month and $53 in the fifth month.

4. How many dollars did Jesse save:
   a) in the first month? _____
   b) in the second month (estimate)? _____

5. In what month did Jesse save:
   a) the most? _____
   b) the least? _____

6. How much more did Jesse save in the fifth month than in the third month? _____

7. In which months were Jesse's savings more than $20? _____

8. In what month were Jesse's savings about twice as much as in the first month? _____

9. Estimate Jesse's total savings for the first 5 months. _____

Build a Vertical Bar Graph

Follow the instructions below, using the information from the table to build a vertical bar graph.

1. Write the title of the graph above the graph.

2. Label the vertical scale with these median ages: 25, 30, 35, 40, and 45. Try to make them equally spaced.

3. Label the horizontal scale with the years that are given. These should be equally spaced too.

4. Using the information from the table, draw a bar for each year that represents the correct median age. The bars should be the same width with equal space between the bars.

United States Median Age	
Year	**Age**
1960	29
1970	27
1980	30
1990	33
2000	36 *

*Predicted age.
Source: U.S. Bureau of the Census.

Reading a Double-Bar Graph

To compare two sets of related information (data), we can use a double-bar graph. The following graph compares daytime sales to evening sales.

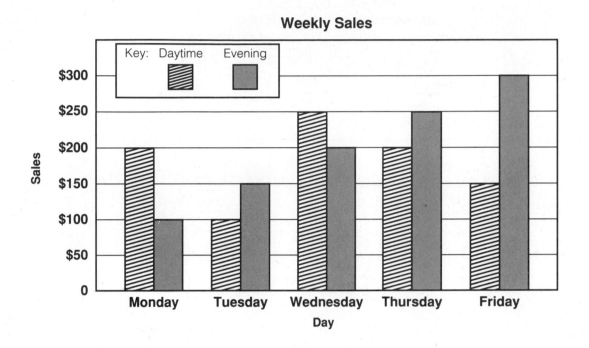

Weekly Sales

▶ Use the double-bar graph to answer the following questions.

1. On what day were evening sales:

    a) the highest? _____

    b) the lowest? _____

2. What was the difference between daytime and evening sales each day?

    a) Monday _____

    b) Tuesday _____

    c) Wednesday _____

    d) Thursday _____

    e) Friday _____

3. What day had the highest daytime sales? _____

4. What was the total evening weekly sales? _____

5. What day had the lowest total of daytime and evening sales? _____

Build a Double-Bar Graph

Information from tables can be visually shown in graph form.

**Participation in Sports
at Haines Junior High School**

Sport	Number of Boys	Number of Girls
Basketball	12	19
Soccer	30	36
Track	38	42
Cross-country	21	7

▶ Use the information from the table above to complete the following double-bar graph.

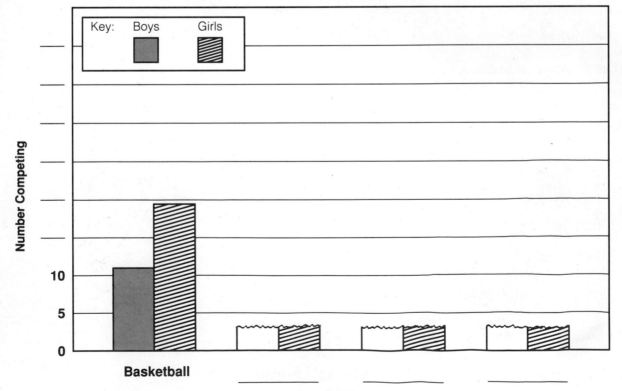

Participation in Sports at Haines Junior High School

Reading a Line Graph

A **line graph** is a graph that uses lines to show patterns, or trends. It is very useful when comparing changes in amounts.

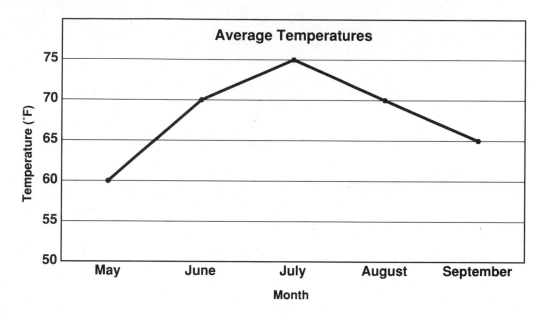

Example: What is the normal temperature for May?

Step 1	Step 2	Step 3
Find May on the horizontal scale.	Find the point on the line graph above May.	Find the temperature on the vertical scale that lines up with the point.

The normal temperature for May is 60°.

▶ Use the graph to answer the following questions.

1. List the normal temperature for each month:

   a) June _____   **c)** August _____

   b) July _____   **d)** September _____

2. What month has the highest normal temperature? _____

3. What month has the lowest normal temperature? _____

4. What 2 months have the same normal temperature? _____

5. State 1 general trend, or pattern, that you notice from the graph. _____

_____

Find the Values

Number of Cases of Flu in Kalamazoo

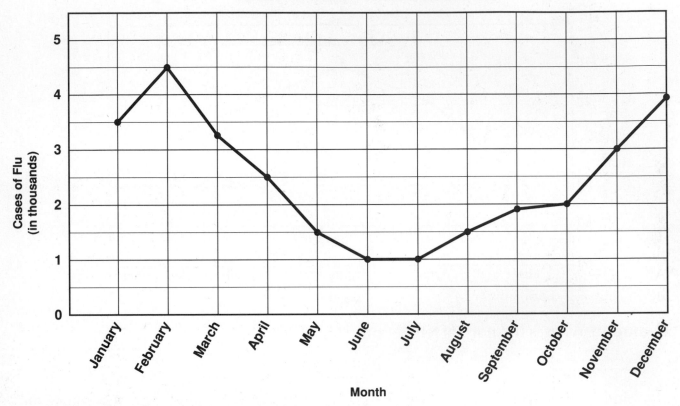

▶ Use the line graph to answer the following questions.

1. List the number of cases of flu (in thousands) for the following months. Use the numbers on the vertical scale to help you find values for points that fall between lines.

   a) January _____   **d)** April _____   **g)** July _____

   b) February _____   **e)** May _____   **h)** August _____

   c) March _____   **f)** June _____   **i)** September _____

2. In which month was the number of cases of flu the greatest? _____

3. In which months was the number of cases of flu the smallest? _____

4. Did the number of cases of flu increase or decrease between July and November? _____

5. Did the number of cases of flu increase or decrease between February and June? _____

Plot the Points

A **line graph** shows changing amounts and general trends.

Points Scored by Lincoln's Best Basketball Player			
Game	Points Scored	Game	Points Scored
1	20	6	40
2	30	7	30
3	35	8	30
4	35	9	20
5	50	10	25

▶ Use the information from the table to complete the line graph. First, finish labeling the horizontal and vertical scales for the line graph.

Then plot the points scored for each game. Sometimes a point is plotted between the horizontal grid lines. When this happens, estimate where the point should be placed. After you have plotted the points, connect them with line segments. Games 1 and 2 have been done for you.

1.

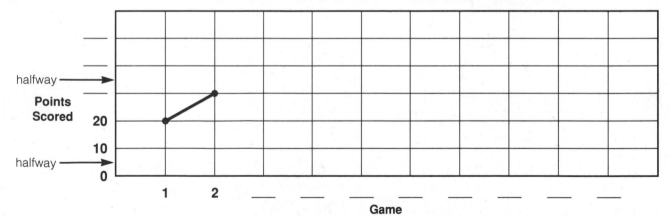

Points Scored By Lincoln's Best Basketball Player

▶ Use the table and line graph to answer the following questions.

2. In what game did he score the most points? _____

3. What was the least number of points that he scored in those 10 games? _____

4. In how many games did he score over 30 points? _____

5. In how many games did he score under 30 points? _____

6. In 3 different games, he scored the same number of points. How many points were scored in each of these games? _____

Practice Helps

Jack owns his own air conditioner repair service. The table below shows how many service calls Jack went on each week for 10 weeks.

Jack's Repair Service			
Week	Number of Calls	Week	Number of Calls
1	8	6	12
2	12	7	14
3	14	8	10
4	17	9	7
5	18	10	4

▶ Using the information from the table, follow the directions to complete the line graph.

- Write the title of the graph.

- Finish labeling the horizontal and vertical scales.

- Using the information from the table, plot and connect the points on the graph.

1.

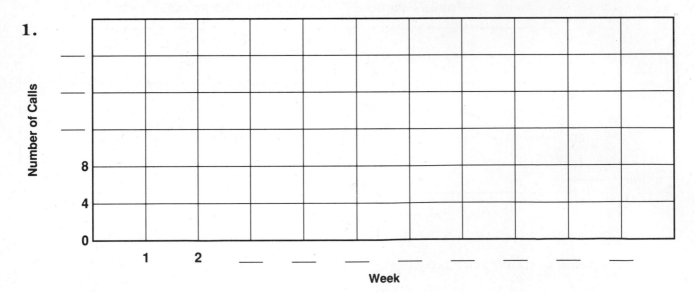

▶ Use the graph to answer the following questions.

2. How many weeks did he go on 10 service calls or more? _____

3. How many weeks did he go on less than 15 service calls? _____

4. Which week was the busiest? _____

5. In which week did he go on the least number of calls? _____

Showing General Trends

▶ Use the table to complete the line graph.

- Finish labeling the horizontal and vertical scales for the graph.

- Plot the point on the graph for each day. Then connect the points with line segments. Days 1, 2, and 3 have been done for you.

Changes in Average Daily Temperature			
Day	Temperature	Day	Temperature
1	20°	6	18°
2	30°	7	41°
3	35°	8	41°
4	20°	9	33°
5	15°	10	25°

1.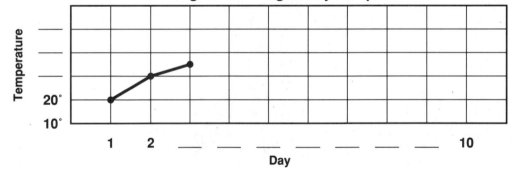

▶ Use the table and line graph to answer the following questions.

2. On how many days was the average temperature above 30°? _____

3. On how many days was the average temperature below 40°? _____

4. On how many days was the average temperature the same as the day before? _____

5. At what time of the year do you think these temperatures were taken? _____

6. On what day did the temperature increase the most over the temperature of the day before? _____

7. What was the range in temperatures? (Range is the difference between the least and greatest measure.) _____

Reading a Double-Line Graph

A double-line graph compares changes in two sets of data over the same period of time.

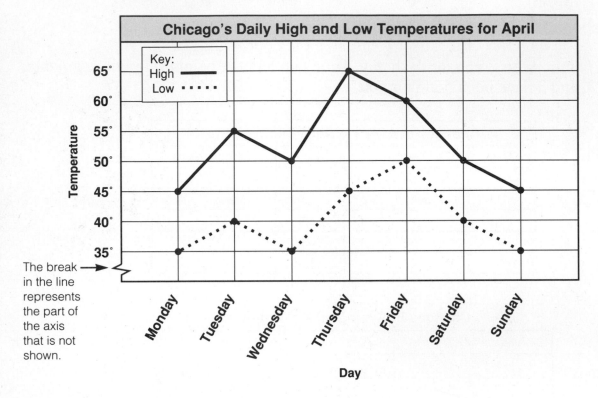

The break in the line represents the part of the axis that is not shown.

▶ Use the graph to answer the following questions.

1. a) On what day was the highest temperature recorded? _____

   b) What was the high temperature for that day? _____

2. On how many days was the temperature over 50°? _____

3. On how many days was the temperature less than 45°? _____

4. Did the day with the highest temperature also have the highest low? _____

5. Were there any days that did not have a temperature increase of at least 10°?

   _____

6. On what days was the difference between the daily high and low temperatures 15°?

   _____

7. a) On what day was the difference between the daily high and low temperatures the greatest? _____

   b) How many degrees' difference was there? _____

Build a Double-Line Graph

Double-line graphs allow you to compare changes in data within a line and between lines.

▶ Use the information from the table to complete the double-line graph.

Books Sold					
Subject	Monday	Tuesday	Wednesday	Thursday	Friday
Math	30	13	54	38	51
English	12	41	22	19	40

1.

Books Sold

Key:
Math ——
English ·····

Books

20
10
0

Monday   Tuesday   ____   ____   ____

Day

▶ Use the graph to answer the following questions.

2. On how many days were more math books sold than English books? _____

3. About how many more English books than math books were sold on Tuesday? _____

4. On what 2 days were the most math books sold? _____

5. On what day were the most books sold? _____

6. By looking at the graph, estimate whether more English or more math books sold during the week. _____

Graph the Data

A line graph shows changing amounts and general trends. Sometimes an amount on a graph falls between labeled amounts. Use those labeled amounts to help you locate the amount that is not labeled.

Hours Spent Watching Television					
Family	Wednesday	Thursday	Friday	Saturday	Sunday
Parents	1	$\frac{1}{2}$	3	$4\frac{1}{4}$	5
Children	3	$2\frac{3}{4}$	1	5	6

▶ Use the information from the table above to complete the double-line graph.

1.

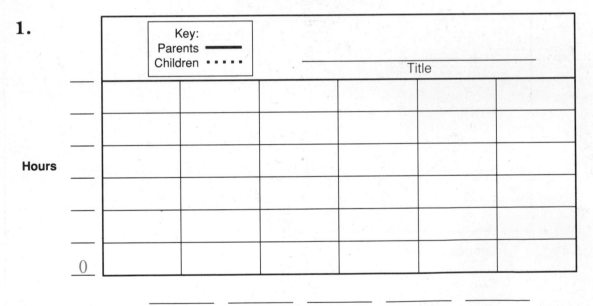

▶ Use the graph to answer the following questions.

2. On what day did parents watch more hours of television than children? _____

_____

3. On what day did children watch the least television? _____

4. Do parents and children seem to watch about the same amount of television? _____

5. Write down 1 trend you notice when looking at the graph. _____

_____

Estimate the Distance

Two things that affect how long it takes to stop a moving car:

- The reaction-time distance (the distance traveled while the driver is reacting to seeing danger and preparing to apply the brakes);

- The total stopping distance (the distance traveled after the driver has applied the brakes).

▶ Use the double-line graph to answer the following questions.

1. Estimate the total stopping distances for the following car speeds.

   a) 20 mph ___45 feet___

   b) 30 mph _____

   c) 40 mph _____

   d) 50 mph _____

   e) 60 mph _____

   f) 70 mph _____

2. Estimate the reaction-time distances for each car speed.

   a) 20 mph _____

   b) 30 mph _____

   c) 40 mph _____

   d) 50 mph _____

   e) 60 mph _____

   f) 70 mph _____

3. What does the graph tell you about the relationship between speed and total stopping distance? _____
_____
_____
_____

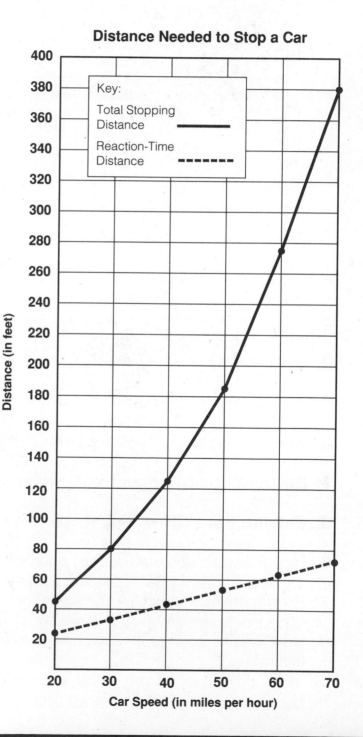

Distance Needed to Stop a Car

Key:

Total Stopping Distance ———

Reaction-Time Distance - - - - -

Distance (in feet)

Car Speed (in miles per hour)

Reading a Tally Table

Sometimes we use tally marks instead of numbers to help us count.

| One tally mark stands for 1 of something.

|||| Four tally marks stand for 4.

卌   The fifth tally mark goes across the first four to make it easy to count groups of 5.

Milk Shake Sales	
Flavor	**Number Sold**
Vanilla	卌 卌 卌 卌 |||
Chocolate	卌 卌	
Strawberry	卌 卌 卌	
Banana	卌 |||

▶ Use the tally table to answer the following questions.

1. What were the 2 most popular flavors? _____

2. What was the least popular flavor? _____

▶ Count by fives to help you answer these questions.

3. How many vanilla milk shakes were sold? _____

4. How many chocolate milk shakes were sold? _____

5. How many strawberry milk shakes were sold? _____

6. How many banana milk shakes were sold? _____

7. How many milk shakes were sold altogether? _____

8. How many more vanilla and chocolate shakes were sold than strawberry and banana? _____

9. How many fewer banana shakes were sold than strawberry? _____

Frequency Tables

Sometimes it is helpful to use tally marks to organize data into a frequency table. The **frequency** of an event is the number of times the event occurs.

30 Students' Test Scores					
7	5	9	7	10	7
6	8	10	8	7	9
7	8	7	9	6	7
10	7	5	6	9	6
7	6	3	7	6	7

▶ Use the test scores above to complete the frequency table.

For each number you cross out, place one tally mark beside the score in the table. After all of the numbers are crossed out, find the frequency for each score. The tens have been done for you.

Since a score of 10 was recorded 3 times, we say the frequency of 10 is 3.

1.

Score	Tally	Frequency
10					3
9		
8		
7		
6		
5		
4		
3		

2. What score occurred the most times? _____

3. The score of 6 had a frequency of _____ .

4. How many different scores were recorded in the frequency table? _____

5. What 3 scores had the highest frequency? _____

Histogram

In a survey, students were asked to name their favorite sandwiches. Find the frequency of each sandwich, and fill in the frequency table.

1.

Sandwich	Tally	Frequency
Ham	ЖЛ	**a)** 5
Roast beef	ЖЛ ||||	**b)**
Chicken	ЖЛ ЖЛ ЖЛ ||	**c)**
Hamburger	ЖЛ ЖЛ ЖЛ ЖЛ	**d)**
Submarine	ЖЛ ЖЛ ||	**e)**

The data from a frequency table can be represented by a graph called a histogram. A **histogram** is similar to a bar graph and shows frequency data visually.

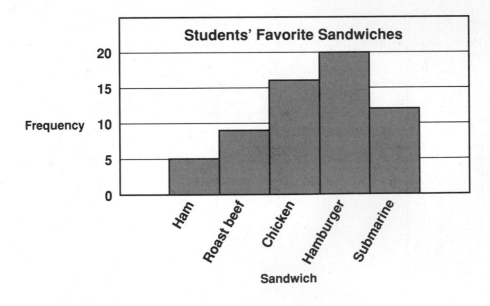

▶ Use the frequency table and histogram to answer the following questions.

2. From the survey, what sandwich was:

   a) most popular? _____   **b)** least popular? _____

3. What was the difference in frequency between the chicken and ham sandwiches? _____

4. What 3 sandwiches had the highest frequency? _____

_____

Working with a Histogram

The weights of players on a local football team were recorded using a frequency table.

1. Find the frequency of each weight interval.

Weight Interval	Tally	Frequency
111–130	||	**a)**
131–150	⊮ ||	**b)**
151–170	⊮ ⊮ ⊮ ||	**c)**
171–190	⊮ |||	**d)**
191–210	⊮ ⊮ ||	**e)**
211–230	|||||	**f)**

 Using the information from the frequency table, complete the histogram below.

2.

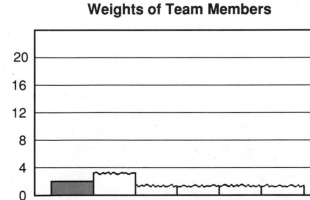

Weights of Team Members

3. What weight interval has the greatest frequency? _____

4. How many more weights were recorded in the 191–210 interval than the 131–150 interval? _____

5. How many players had weights greater than 150 pounds? _____

6. When looking at the histogram, what trend do you see? _____

_____

Grouping Data Using a Histogram

A survey listed the ages of workers at a local supermarket. The results were as follows:

~~18~~	~~47~~	38	34	29	36	34	45	60	45
43	39	24	18	55	49	24	18	18	31
26	25	30	23	19	48	22	29	27	35
42	20	32	38	41	35	39	26	37	24

▶ Use the results to organize the information in the following frequency table. First record a tally mark for each age in the row for the correct interval. Cross out each number after recording the tally mark. Then fill in the frequency for each age interval.

Age Interval	Tally	Frequency
18–25	1. a) |	b)
26–32	2. a)	b)
33–39	3. a)	b)
40–46	4. a)	b)
47–53	5. a) |	b)
54–60	6. a)	b)

7. Using the information from the frequency table, complete the histogram.

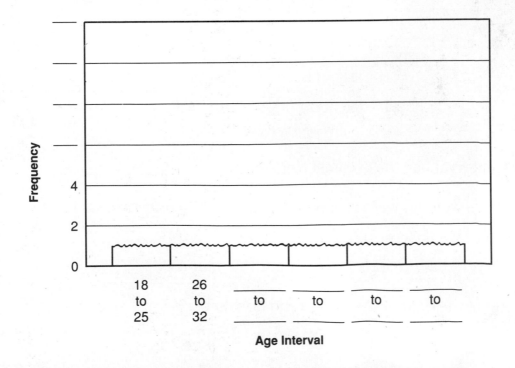

Line Plots

Line plots are a quick, simple way to organize data.

The following data list the total number of hours each student watched television for a week.

$$\cancel{3} \quad \cancel{12} \quad 15 \quad 10 \quad 17 \quad 16 \quad 17 \quad 15 \quad 13 \quad 12$$
$$23 \quad 15 \quad 17 \quad 11 \quad 13 \quad 16 \quad 18 \quad 19 \quad 15 \quad 11$$

▶ To make a line plot of the data, first draw a number line long enough to include the range of numbers. Then, for each value, mark an ✕ at the point on the number line that matches the value. To represent the first number, 3, place an ✕ above the line at 3. To represent the second number, 12, place an ✕ above the line at 12. Cross out each number after recording it on the plot.

1. Continue placing ✕s over the number line to complete the line plot below.

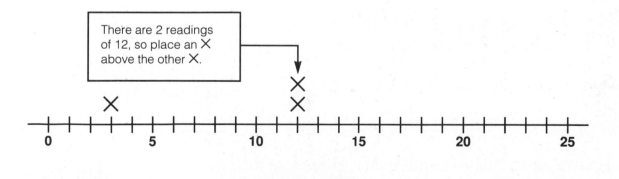

There are 2 readings of 12, so place an ✕ above the other ✕.

▶ Using the line plot, answer the questions below.

2. How many students watched television for 15 hours? _____

3. How many students watched television for 8 to 20 hours? _____

4. How many students watched television for more than 14 hours? _____

5. Over which value are the most ✕s gathered? _____

Plot the Numbers

The following data list the ages of each student in class:

19	35	25	22	39
33	20	25	30	22
18	23	28	18	20
24	22	19	20	25

1. Complete the line plot by placing ✕s above the number line for each number listed above. Remember to cross off each number from the list after you use it.

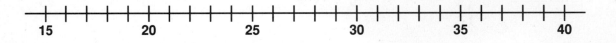

▶ Using the line plot, answer the questions below.

2. How many students are less than 30 years old? _____

3. How many students are more than 25 years old? _____

4. How many students are more than 18 but less than 24 years of age? _____

5. The 4 oldest students in class are how old? _____

6. List any trends that you see shown on the line plot. _____

_____

_____

_____

10 MEAN, OR AVERAGE

Learning about Averages

The **mean,** or average, is a good measure for describing the middle amount of a collection of data. To find the mean, or average, of a set of data, divide the total by the number of parts.

Example: Diane's test scores were 95, 75, 84, and 70.
What was her average score?

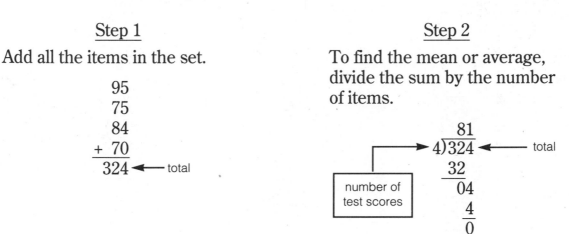

Step 1

Add all the items in the set.

```
  95
  75
  84
+ 70
 324  ← total
```

Step 2

To find the mean or average, divide the sum by the number of items.

```
       81
     4)324  ← total
       32
       04
        4
        0
```

number of test scores

The mean, or average, score was 81.

▶ Average the following test scores, and use the grading scale to assign a grade for each student.

Student	Test Scores							Total Score	Average Score	Grade
Todd	85	70	82					1. a)	b)	c)
Elva	70	84	96	60	75			2. a)	b)	c)
Daisy	88	95	100	89	91	100	88	3. a)	b)	c)
Doug	70	54	81	43				4. a)	b)	c)
Beth	100	70	69	89	92			5. a)	b)	c)
Tiffany	71	82	90	54	88	95		6. a)	b)	c)

Grading Scale

100–90 = A	89–80 = B	79–70 = C	69–60 = D	59 or less = E

7. What student had the highest grade? _____

8. What student had the lowest grade? _____

Find the Mean

Remember, to find the mean, or average, of a set of data, divide the total by the number of parts.

Example: What was the average rainfall?

Monthly Rainfall for 3 months
May: 5.5 inches
June: 2.4 Inches
July: 6.8 inches

Step 1
Add the numbers.

```
   5.5
   2.4
 + 6.8
 ─────
  14.7
```

Step 2
Divide by the number of months.

```
        4.9
  3)14.7  ◄── total inches
     12
     ──
     2 7
     2 7
     ───
       0
```
number of months

The average rainfall was 4.9 inches.

___

▶ Find the mean, or average, in each problem. Add the numbers, then divide by the number of items.

1. In the first 4 games, Lori scored as follows:

Game 1:   15 points
Game 2:   9 points
Game 3:   23 points
Game 4:   13 points

What was her average score, stated as points per game? _____

2. The last 5 telephone bills were for the following amounts:

Bill 1: $12.38
Bill 2: $17.25
Bill 3: $14.92
Bill 4: $25.10
Bill 5: $30.05

What was the average amount of the telephone bills? _____

3. Kim went on a 4-day vacation. He traveled the following distances:

First day: 463 miles
Second day: 254 miles
Third day: 135 miles
Fourth day: 932 miles

What was the average distance he traveled in a day? _____

4. A salesperson earned the following commissions in the past 5 weeks:

Week 1: $295
Week 2: $183
Week 3: $379
Week 4: $488
Week 5: $175

What was the salesperson's average weekly commission? _____

Find the Average

Sometimes you must find the average when the total amount is given.

Example

Trish played in 21 basketball games and scored a total of 294 points. What was Trish's game average?

To find the average, divide the total points scored by the number of games.

$$\begin{array}{r} 14 \\ 21\overline{)294} \\ \underline{21} \\ 84 \\ \underline{84} \\ 0 \end{array}$$

number of games ← → total points scored

Trish's game average was 14 points.

▶ Find the average in each problem.

1. Loren traveled 228 miles using 12 gallons of gas. What was his average mileage for 1 gallon of gas?

Loren traveled _____ miles on 1 gallon of gas.

3. The repair bills for 9 cars came to $405. What was the average repair bill per car?

The average repair bill per car was

_____ .

2. A class of 28 students sold 112 magazine subscriptions. What was the average number of subscriptions sold per student?

The average number of subscriptions sold per student was _____.

4. Dee traveled 384 miles in 8 hours. What was her average speed per hour?

Dee's average speed was _____ miles per hour.

Averaging Sales

The average, or mean, is commonly used in business—for example, to find average sales over some period of time. Remember, to find the average of a group of numbers, add all the numbers together, then divide the sum by the number of items.

▶ Use the book sales figures for the month of June to find the total and average sales for each week.

1. Week 1:
$8,000	$6,000	$4,000	$10,000	$3,000
Monday	Tuesday	Wednesday	Thursday	Friday

    a) Total sales: _____   **b)** Average sales: _____

2. Week 2:
$19,000	$9,000	$15,000	$11,000	$22,000
Monday	Tuesday	Wednesday	Thursday	Friday

    a) Total sales: _____   **b)** Average sales: _____

3. Week 3:
$13,000	$4,000	$6,000	$6,000	$7,000
Monday	Tuesday	Wednesday	Thursday	Friday

    a) Total sales: _____   **b)** Average sales: _____

4. Week 4:
$10,000	$29,000	$18,000	$17,000	$16,000
Monday	Tuesday	Wednesday	Thursday	Friday

    a) Total sales: _____   **b)** Average sales: _____

▶ Use the total and average book sales in questions 1–4 to complete the table.

June Book Sales		
Week	**Total Sales**	**Average Sales**

Graph the Sales

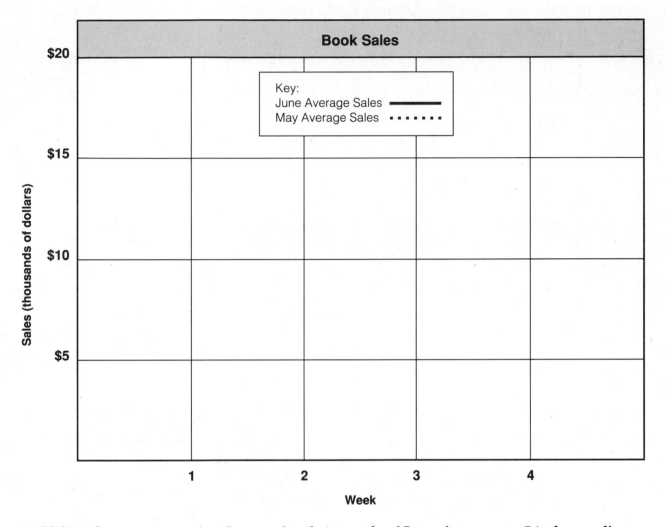

Book Sales

Key:
June Average Sales ———
May Average Sales · · · · · ·

Sales (thousands of dollars)

$20
$15
$10
$5

Week
1   2   3   4

1. Using the average sales figures for the month of June from page 54, draw a line graph.

2. For the month of May, the average sales figures were as follows:

   Week 1: $13,400   Week 2: $7,900   Week 3: $10,100   Week 4: $6,900

▶ Using the average sales figures for the month of May, complete the graph above by making a double-line graph. Use the key as a guide.

3. Which month had the higher sales figure for:

   a) Week 1?_____   **c)** Week 3?_____

   b) Week 2?_____   **d)** Week 4?_____

4. Which month had the greater sales overall? _____

Rank Order

In organizing data, it is sometimes necessary to rank the numbers in order of size.

▶ Rank the following sets of numbers in the order required.

1. Eric's test scores:

| 75 | 90 | 100 | 82 | 68 | 95 | 72 |

_____ _____ _____ _____ _____ _____ _____
lowest   highest

2. Height in feet of the 5 tallest buildings in Chicago:

| 859 | 1,454 | 1,136 | 850 | 1,127 |

_____ _____ _____ _____ _____
shortest   tallest

3. Corn production in bushels:

| 232,400 | 367,080 | 632,000 | 172,540 | 104,760 |

_____ _____ _____ _____ _____
least   most

4. Size of 4-year colleges in terms of number of students:

| 1,906 | 742 | 10,222 | 4,252 | 26,963 |

_____ _____ _____ _____ _____
smallest   largest

5. Average high temperatures:

| 86° | 75° | 92° | 79° | 90° | 78° |

_____ _____ _____ _____ _____ _____
lowest   highest

Median and Mode

There are other ways to judge the values of a series of numbers. The **median** is the middle value in a set of numbers. The **mode** is the number that appears most often.

Example: A softball team scores runs of 5, 11, 7, 12, 9, 5, and 4 in seven games. To find the median of the runs scored, rank the numbers from smallest to largest.

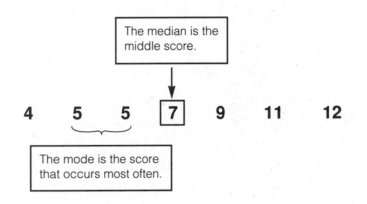

The median is the middle score.

4   5   5   7   9   11   12

The mode is the score that occurs most often.

▶ Arrange each set of data from smallest to largest. Then find the median and mode. Some sets may not have a mode; if so, write the word *none*.

		Median	Mode
1.	2, 6, 9, 4, 7, 6, 3	_____	_____
	2, 3, 4, 6, 6, 7, 9		
	smallest largest		
2.	19, 14, 11, 15, 19	_____	_____
3.	107, 119, 138, 192, 148	_____	_____
4.	38, 17, 24, 85, 17, 95, 62, 35, 20	_____	_____
5.	7, 9, 12, 15, 8, 7, 7, 5, 9	_____	_____

Find the Median

The median is the middle value in a set of numbers. If a set of values contains an even number (2, 4, 6, 8, . . .) of items, the median is the average (mean) of the two middle numbers.

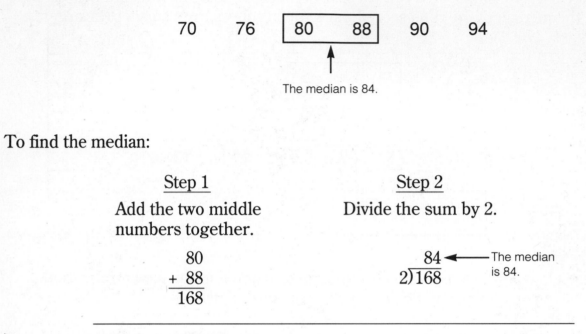

70   76 | 80   88 | 90   94

The median is 84.

To find the median:

Step 1	Step 2
Add the two middle numbers together.	Divide the sum by 2.

Step 1:
$$\begin{array}{r} 80 \\ +\ 88 \\ \hline 168 \end{array}$$

Step 2:
$$2\overline{)168} = 84 \quad \longleftarrow \text{The median is 84.}$$

▶ Arrange each set of data from smallest to largest. Then find the median.

Median

1.   62   51   68   58

     51 | 58   62 | 68
     smallest   largest

_____

2.   15   19   12   11   20   23

_____

3.   15   9   25   11   17   32   29   23

_____

4.   251   132   72   120   145   140

_____

5.   98   87   55   79   72   101   68   75

_____

Plot the Median

The median of a set of numbers is the middle number once the numbers are arranged in order of size.

1. Find the median of Dana's daily class scores in each of her classes.

Class	Dana's Class Scores							Median
English	94	63	80	100	78	92	96	**a)**
Math	80	85	90	68	87	93	55	**b)**
Science	78	66	92	71	52	69	83	**c)**
History	77	85	70	100	80	45	68	**d)**
Business	100	70	98	85	100	99	78	**e)**

2. Use Dana's median class scores from the table to complete the following bar graph.

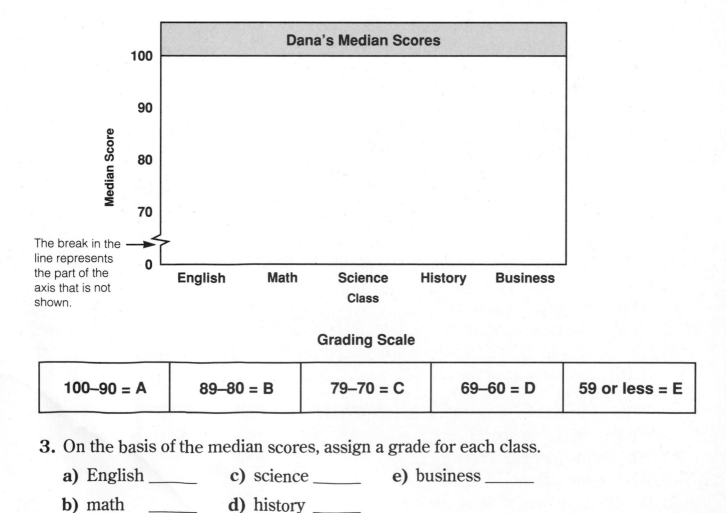

Grading Scale

100–90 = A	89–80 = B	79–70 = C	69–60 = D	59 or less = E

3. On the basis of the median scores, assign a grade for each class.

a) English _____   **c)** science _____   **e)** business _____

b) math _____   **d)** history _____

Mean, Median, and Range

Finding the mean, median, and range helps to organize data to understand it more clearly. Remember these definitions:

- The **mean** of a set of numbers is the sum of the numbers divided by the number of items.

- The **median** of a set of numbers is the middle number once the numbers have been arranged in order of size.

- The **range** in a set of numbers is the difference between the largest and smallest numbers.

▶ Answer the questions about the following sets of data.

1. The high temperatures for the first 5 days in July were 62°, 86°, 73°, 88°, and 81°.
   a) What is the mean temperature? _____
   b) What is the median temperature? _____
   c) What is the range in temperatures? _____
   d) What temperatures are below the median? _____
   e) What temperatures are greater than the mean? _____

2. Tamala's scores on the math quizzes were 85, 65, 87, 68, 90, and 73.
   a) What is the mean score? _____
   b) What is the median score? _____
   c) What is the range? _____
   d) What scores are below the median? _____
   e) What scores are greater than the mean? _____

3. Steven kept a record of his savings for 5 weeks. He saved $5, $15, $20, $30, and $80.
   a) What is the mean savings? _____
   b) What is the median savings? _____
   c) What is the range in savings? _____
   d) If Steven saved $150 instead of $80 in the 5th week, would this cause a greater change in the mean or the median? _____

4. Gigi's bowling scores were 183, 123, 116, 168, 142, and 150.
   a) What is the mean score? _____
   b) What is the median score? _____
   c) What is the range? _____
   d) How many scores were above the mean? _____

Using Data

Month	April	May	June	July	August	September
Number of Rainy Days	13	6	3	5	4	5
Number of Sunny Days	10	12	15	23	19	17

▶ Use the chart to answer the following questions.

1. What is the total number of rainy days? _____

2. What is the total number of sunny days? _____

3. What is the mean for:
   a) rainy days? _____
   b) sunny days? _____

4. What is the median for:
   a) rainy days? _____
   b) sunny days? _____

5. What is the range for:
   a) rainy days? _____
   b) sunny days? _____

6. What is the mode for:
   a) rainy days? _____
   b) sunny days? _____

7. Use a double-bar or double-line graph to graph the data above.

8. Based on the graph, write two statements about any comparisons you can make or trends you see.

   a) _____

   b) _____

Finding Data

The average is a good measure for describing the middle amount of a collection of data.

Name	Bowling Scores				
Joel	110	125	85	143	121
Lori	158	138	160	125	164
Anna	128	108	110	134	105
Martin	149	137	152	139	143
Ian	205	173	183	159	160
Scott	141	137	127	131	118
Donna	204	183	205	192	210

▶ Using the data above, find the average score, high score, and low score for each bowler. Some average scores may need to be rounded. Complete the following table.

Name	Average Score	High Score	Low Score
Joel	1. a)	b)	c)
Lori	2. a)	b)	c)
Anna	3. a)	b)	c)
Martin	4. a)	b)	c)
Ian	5. a)	b)	c)
Scott	6. a)	b)	c)
Donna	7. a)	b)	c)

Graph the Data

1. Using the average, high, and low scores you found, complete the line graph showing the scores for each bowler as shown on the chart on page 62.

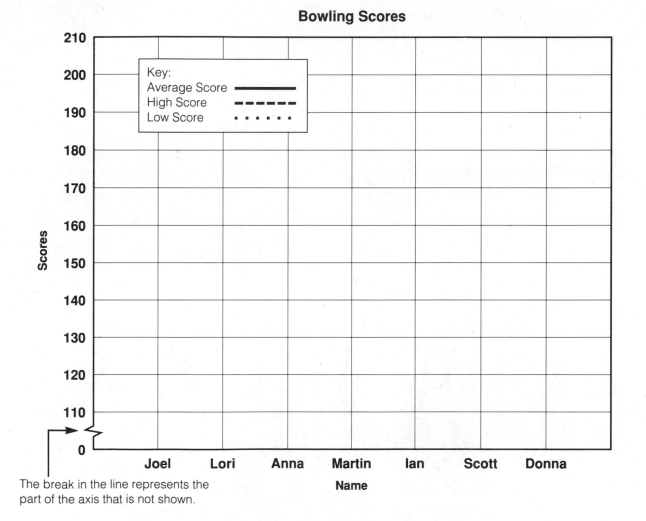

The break in the line represents the part of the axis that is not shown.

▶ Use the data from the table or graph to answer the following questions.

2. Which bowler had the highest average score? _____

3. Which 2 bowlers had the lowest scores? _____

4. What was the difference between Scott's high and low scores? _____

5. Which bowler's high score was the most different from his or her average? _____

6. Rank the bowlers from the highest to the lowest average score.

Donna   _____   _____   _____   _____   _____   _____
highest   lowest

Comparing Graphs

- A line graph shows trends.
- A bar graph compares similar categories.
- A circle graph compares the sizes of the parts of a whole.

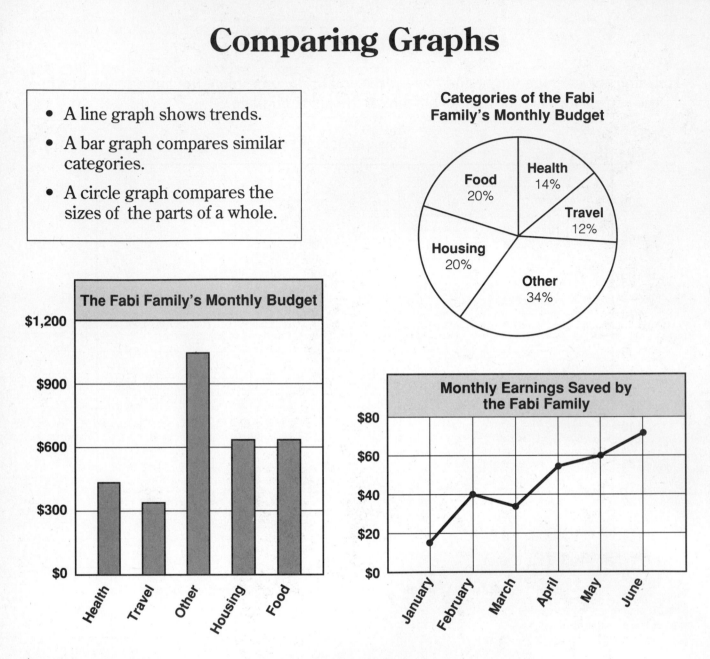

Categories of the Fabi Family's Monthly Budget

Food 20%
Health 14%
Travel 12%
Housing 20%
Other 34%

The Fabi Family's Monthly Budget

$1,200
$900
$600
$300
$0

Health · Travel · Other · Housing · Food

Monthly Earnings Saved by the Fabi Family

$80
$60
$40
$20
$0

January · February · March · April · May · June

▶ Use the graphs to answer the following questions.

1. Which categories of the Fabis' budget were between $300 and $600? _____
   _____

2. Travel made up what percent of the budget? _____

3. Which graph shows how much the Fabi family saved? _____

4. a) What 2 expenses total about $1,200? _____
   b) About what percent of the budget is this? _____

5. How much money did the Fabi family save in June? _____

Practice Your Skills

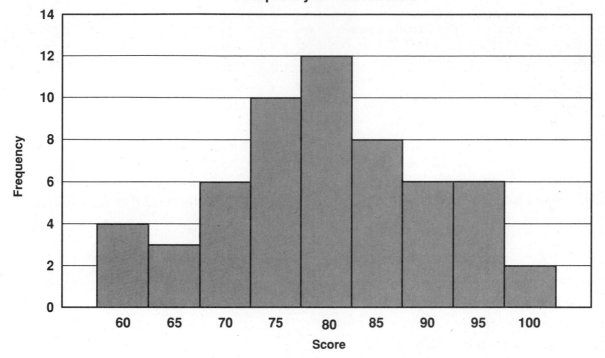

Frequency of Test Scores

▶ Use the histogram to answer the following questions.

1. a) What score shows a "frequency of 10?" _____

    b) What does this mean? _____

2. List the frequency of each score.

Score	60	65	70	75	80	85	90	95	100
Frequency									

3. How many students took the test? _____

4. Which score had the lowest frequency? _____

5. What is the range of scores? _____

6. State 1 fact that you notice about the histogram. _____
_____
_____

GRAPHS REVIEW

▶ Use the graphs to answer the questions.

Family Budget

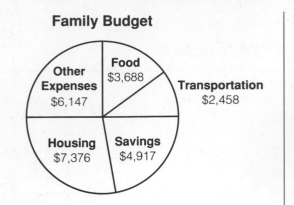

Runs Scored by Tigers

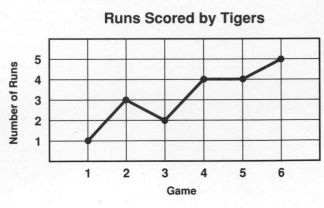

1. What is the title? _____

2. What budget item is $3,688? _____

3. What is the largest budget item? _____

4. How much money was budgeted for housing and transportation? _____

9. In what game did the team score:

   a) the most runs? _____

   b) the fewest runs? _____

10. Do you see a general trend? _____ What trend do you see? _____ _____

Education and Total Earnings

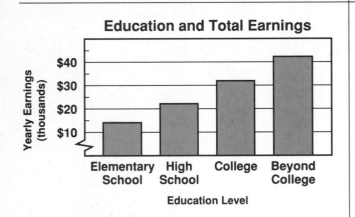

Daily Sales

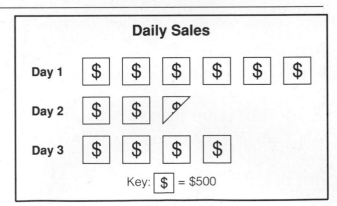

Key: $ = $500

5. The title of the graph is_____ _____

6. The vertical scale shows _____ _____

7. The horizontal scale shows _____ _____

8. Estimate the yearly earnings for each education level.

   a) elementary school _____

   c) college _____

   b) high school _____

   d) beyond college _____

11. Each $ represents how much money? _____

12. What are the sales for:

   a) Day 1? _____

   b) Day 2? _____

   c) Day 3? _____

13. The combined sales of Days 2 and 3 are how much more than the sales of Day 1? _____

14. What are the total sales for the 3 days? _____

ANSWER KEY

Page 1: Reading Tables

1. 4
2. 40
3. 30
4. Buildings A, B, and D
5. Building C

Page 2: Calendars Are Tables

1. Thursday
2. July 13
3. July 26
4. 5
5.

6. Wednesday
7. August 23
8. August 27

August						
Sun.	Mon.	Tue.	Wed.	Thu.	Fri.	Sat.
				1	2	3
4	5	6	7	8	9	10
11	12	13	14	15	16	17
18	19	20	21	22	23	24
25	26	27	28	29	30	31

Page 3: Working with a Schedule

1. 5
2. 7:35 A.M.
3. 11:18 P.M.
4. 9:58 A.M.
5. 6:51 P.M.
6. 2 hours, 58 minutes

Page 4: Finding Bus Fares

1. $2.25
2. $18.40
3. a) $2.55
  b) $6.70
  c) a round-trip ticket
4. a) $49.75
  b) $48.60
  c) 2 five-trip tickets to Rupert

Page 5: Compare Rates Between Tables

1. a) $11.00   **b)** yes, a 2-inch ad
2. a) $22.00   **b)** a 6-inch ad
3. a) a 2-inch ad   **b)** yes, for $7.50
4. a) 8-inch and 12-inch ads
  b) 6-inch, 8-inch, and 12-inch ads

Page 6: Loan Payment Schedule

1. a) $15.93   **d)** $14.27
  b) $101.09   **e)** $24.15
  c) $8.94   **f)** $67.39
2. a) 12   **b)** 24
3. a) $9.51   **b)** $228.24
4. a) $3,423.12 **b)** $3,639.24

Page 7: More Work with Loan Schedules

1. 3 years
2. 1 year
3. 5 years
4. $800

Page 7: More Work with Loan Schedules (continued)

5. $2,000
6. $2,000

Page 8: Comparing Prices

1. The prices of different products in America and Japan
2. a) Australia
  b) Japan, United States
  c) United States
  d) Japan, United States
  e) Europe
  f) Japan
3. Blue jeans, movie tickets, pizza, bed linens
4. Calculator, tires
5. $4.02

Page 9: Using a Highway Mileage Chart

1. 809
2. 1,467
3. 2,552
4. 1,627 miles
5. 1,672 miles

Page 10: Postage-Rate Table

1. $2.74
2. $2.41
3. $1.61
4. $2.17
5. $3.05
6. $3.14
7. $5.71
8. $7.80

Page 11: Sales Tax Table

1.

Value Hardware		
Item	Price	
1 Paintbrush	$ 3	48
1 Flood lamp	1	88
Thank You	Subtotal	5	36
	Tax		32
	Total	$ 5	68

2.

Value Hardware		
Item	Price	
3 Sponges	$ 2	97
1 Duct tape	2	35
1 Masking tape		59
Thank You	Subtotal	5	91
	Tax		35
	Total	$ 6	26

Page 11: Sales Tax Table (continued)

3.

Value Hardware		
Item	**Price**	
2 Masking tapes	$ 1	18
1 Pair of work gloves	2	85
Thank You — Subtotal	4	03
Thank You — Tax		24
Thank You — Total	$ 4	27

4.

Value Hardware		
Item	**Price**	
2 Duct tapes	$ 4	70
Thank You — Subtotal	4	70
Thank You — Tax		28
Thank You — Total	$ 4	98

Page 12: Comparing Information

1. Favorite forms of exercise
2. 5 million
3. 4
4. Exercising with equipment
5. Walking
6. Aerobics
7. 25 million

Page 13: Finding Information

1. Cost of a pound of apples in selected world capitals
2. 4
3. $.30 per pound
4. $.15
5. Washington, D.C.
6. Pretoria
7. $.30

Page 14: Using Graphs to Make Decisions

1. 20 minutes
2. 25 minutes
3. Thursday
4. 10 minutes
5. a) Wednesday
   b) 10 minutes
6. Friday

Page 15: Graph the Information

1.

Average Time Spent at Events	
Attraction	**Average Length of Activity**
NFL football game	🕐 🕐 🕐
Amusement park	🕐 🕐 🕐 🕐 🕐
Ice hockey game	🕐 🕐 🕐 🕐
Horse races	🕐 🕐 🕐 🕐

Key: 🕐 = 1 hour

Source: International Food Service Manufacturers' Association.

2. Average time spent at events
3. a) With one-half the symbol for 1 hour
   b) 30 minutes
4. 2 hours more
5. .5 or one-half hour more

Page 16: Using Tables and Graphs

1.

Largest Cities in the World by the Year 2000	
City	**Number of People**
New York, USA	🚶 🚶 🚶
Tokyo, Japan	🚶 🚶 🚶 🚶 🚶 🚶
Seoul, South Korea	🚶 🚶 🚶 🚶
Mexico City, Mexico	🚶 🚶 🚶 🚶 🚶

Key: 🚶 = 5,000,000 (5 million) people

Basic Data: U.S. Census Bureau.

2. 5,000,000
3. Tokyo
4. New York
5. Tokyo and Mexico City

Page 17: Working with Circle Graphs

1. 100%
2. 20%
3. $1.00
4. $.25
5. 1
6. 1/4

Page 18: Comparing Circle Graphs

1. Carbohydrates, fat, and saturated fat
2. Carbohydrates, protein, and fat
3. Alcohol
4. Carbohydrates and fat
5. Protein and fat
6. Carbohydrates and fat

Page 19: From Table to Graph

1. **Percent of Mike's Monthly Earnings**

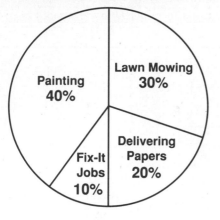

2. Painting
3. Delivering papers
4. 70%
5. $24.00
6. Answers will vary.

Page 20: Show the Percents

1. **Family Budget**

2. One-fourth of the whole circle
3. 60%
4. 4
5. 100%
6. a) Clothing and savings
   b) Yes
7. Answers will vary.

Page 21: Earnings Statement

1. a) 62
   b) 15
   c) 8
   d) 5
   e) 10
2. Yes
3. $62

Page 22: Reading Horizontal Bar Graphs

1. 20
2. 75
3. $2\frac{1}{2}$
4. 100

5. 250
6. 25
7. 9
8. 30

Page 23: Comparing Values

1. a) 80
   b) 50
2. 30
3. a) 45
   b) 15
4. 30
5. a) 40
   b) 10
6. 30
7. a) 80
   b) 40
8. 40

Page 24: Compare the Bars

1. Number of miles 5 students drive in 1 week
2. Todd, Barry, and Rita
3. Todd
4. a) 350
   b) 200
   c) 400
   d) 50
5. Eric
6. 1,250

Page 25: Estimate the Percent

1. Percentage of home gardeners who grow each of 6 vegetables
2. a) about 58%
   b) about 25%
   c) about 85%
   d) 50%
   e) 20%
   f) about 42%
3. Potatoes
4. Peppers, tomatoes, and onions

Page 26: Graphing Monthly Expenses

1. Refer to graph.
2. Refer to graph.
3. Refer to graph.
4. May
5. March
6. $200
7. $50
8. $4,050

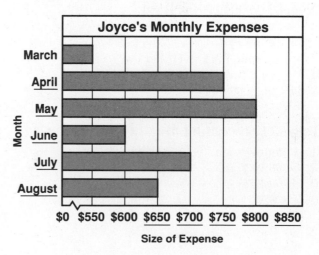

Page 27: Build a Horizontal Bar Graph
1.

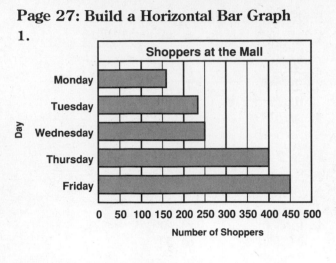

Page 28: Reading Vertical Bar Graphs
1. 20   4. 750
2. 150   5. 75
3. 7   6. 150

Page 29: Comparing Values

1. a) 40   5. a) About 33
   b) 10   b) About 15
2. 30   6. About 18
3. a) 90   7. a) About 90
   b) 30   b) About 25
4. 60   8. About 65

Page 30: Compare the Bars
1. Whitney, Granite Peak, McKinley, and Hood
2. Marcy, Mitchell, and Hood
3. a) McKinley   d) Hood
   b) Whitney   e) Mitchell
   c) Granite Peak   f) Marcy
4. Marcy, Mitchell, and Hood
5. About 5,000 feet

Page 31: Graphing Monthly Savings
1. Refer to graph.   6. About $10
2. Refer to graph.   7. Months 2, 3, and 5
3. Refer to graph.   8. 3
4. a) $20   9. Estimates can range
   b) About $62   from $180–$195.
5. a) 2
   b) 4

Page 31: Graphing Monthly Savings (continued)

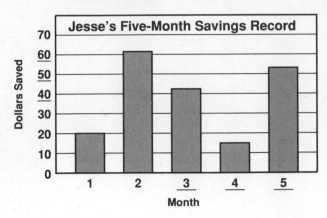

Page 32: Build a Vertical Bar Graph
1.

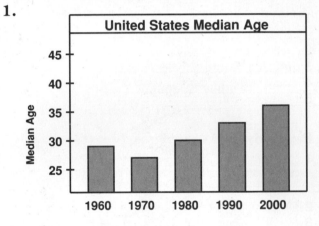

Page 33: Reading a Double-Bar Graph
1. a) Friday   b) Monday
2. a) $100   d) $50
   b) $50   e) $150
   c) $50
3. Wednesday
4. $1,000
5. Tuesday

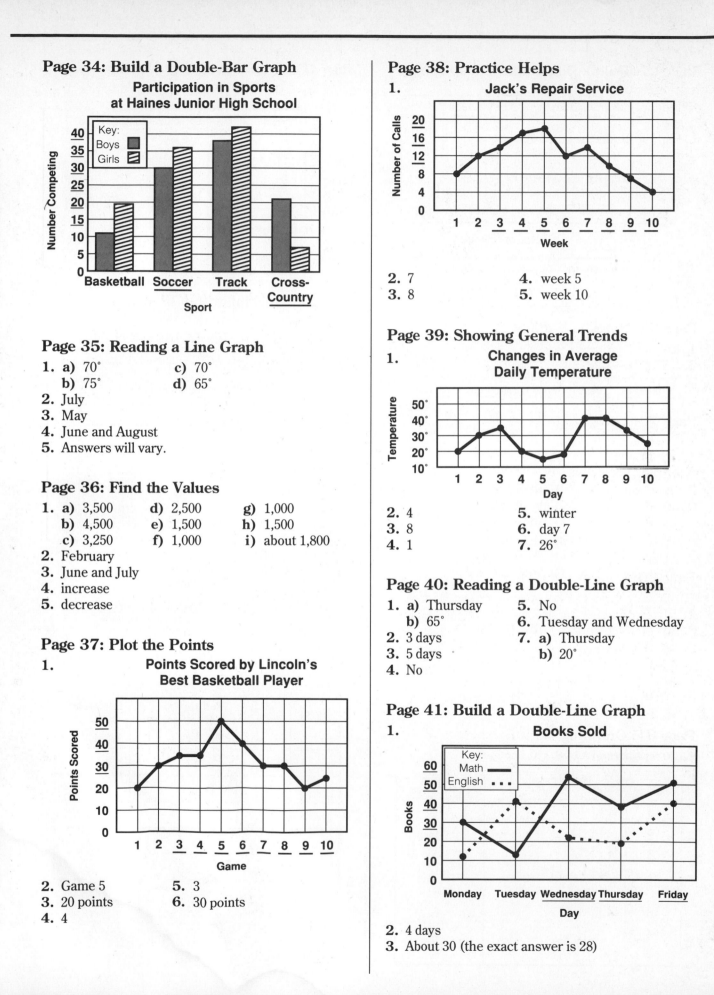

Page 34: Build a Double-Bar Graph

**Participation in Sports
at Haines Junior High School**

Page 35: Reading a Line Graph

1. **a)** 70°   **c)** 70°
   b) 75°   **d)** 65°
2. July
3. May
4. June and August
5. Answers will vary.

Page 36: Find the Values

1. **a)** 3,500   **d)** 2,500   **g)** 1,000
   b) 4,500   **e)** 1,500   **h)** 1,500
   c) 3,250   **f)** 1,000   **i)** about 1,800
2. February
3. June and July
4. increase
5. decrease

Page 37: Plot the Points

1. **Points Scored by Lincoln's
Best Basketball Player**

2. Game 5   5. 3
3. 20 points   6. 30 points
4. 4

Page 38: Practice Helps

1. **Jack's Repair Service**

2. 7   4. week 5
3. 8   5. week 10

Page 39: Showing General Trends

1. **Changes in Average
Daily Temperature**

2. 4   5. winter
3. 8   6. day 7
4. 1   7. 26°

Page 40: Reading a Double-Line Graph

1. **a)** Thursday   5. No
   b) 65°   6. Tuesday and Wednesday
2. 3 days   7. **a)** Thursday
3. 5 days   **b)** 20°
4. No

Page 41: Build a Double-Line Graph

1. **Books Sold**

2. 4 days
3. About 30 (the exact answer is 28)

Page 41: Build a Double-Line Graph (continued)

4. Wednesday and Friday
5. Friday
6. More math books were sold.

Page 42: Graph the Data

1. Refer to graph.

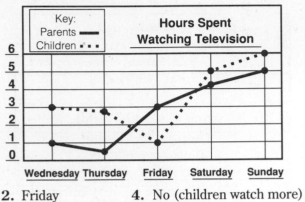

2. Friday
3. Friday
4. No (children watch more)
5. Answers will vary.

Page 43: Estimate the Distance

1. a) About 45 feet   d) About 185 feet
   b) About 80 feet   e) About 275 feet
   c) About 125 feet f) About 380 feet

2. a) About 22 feet   d) About 52 feet
   b) About 32 feet   e) About 62 feet
   c) About 42 feet   f) About 72 feet

3. Answers will vary.

Page 44: Reading a Tally Table

1. Vanilla and strawberry
2. Banana
3. 23
4. 11
5. 16
6. 8
7. 58
8. 10 more
9. 8 fewer

Page 45: Frequency Tables

1.

Score	Tally	Frequency
10					3
9						4
8					3
7	⊞⊞ ⊞⊞		11
6	⊞⊞		6
5				2
4		0
3			1

Page 45: Frequency Tables (continued)

2. 7
3. 6
4. 8
5. 9, 7, and 6

Page 46: Histogram

1. a) 5   d) 20
   b) 9   e) 12
   c) 17

2. a) Hamburger   b) Ham
3. 12
4. Chicken, hamburger, and submarine

Page 47: Working with a Histogram

1. a) 2   d) 8
   b) 7   e) 12
   c) 17 f) 4

2. Refer to graph.

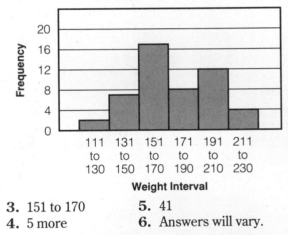

3. 151 to 170
4. 5 more
5. 41
6. Answers will vary.

Page 48: Grouping Data Using a Histogram

Age Interval	Tally	Frequency
18–25	1. a) ⊞⊞ ⊞⊞ ||	b) 12
26–32	2. a) ⊞⊞ |||	b) 8
33–39	3. a) ⊞⊞ ⊞⊞	b) 10
40–46	4. a) ⊞⊞	b) 5
47–53	5. a) |||	b) 3
54–60	6. a) ||	b) 2

Page 48: Grouping Data Using a Histogram (continued)

7. Refer to table.

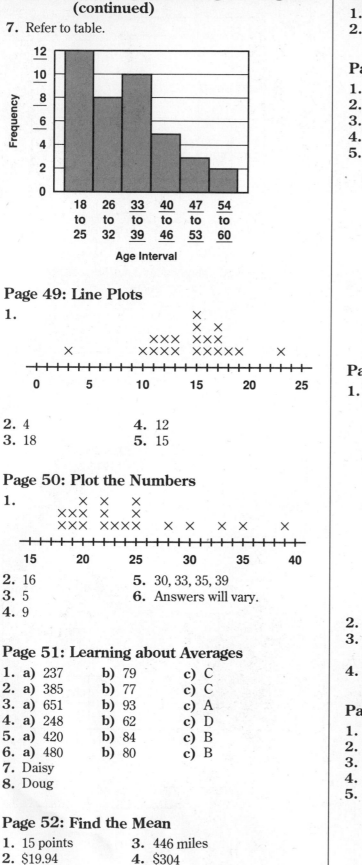

Page 49: Line Plots

1.

(line plot)

2. 4   **4.** 12
3. 18   **5.** 15

Page 50: Plot the Numbers

1.

(line plot)

2. 16   **5.** 30, 33, 35, 39
3. 5   **6.** Answers will vary.
4. 9

Page 51: Learning about Averages

1. a) 237   **b)** 79   **c)** C
2. a) 385   **b)** 77   **c)** C
3. a) 651   **b)** 93   **c)** A
4. a) 248   **b)** 62   **c)** D
5. a) 420   **b)** 84   **c)** B
6. a) 480   **b)** 80   **c)** B
7. Daisy
8. Doug

Page 52: Find the Mean

1. 15 points   **3.** 446 miles
2. $19.94   **4.** $304

Page 53: Find the Average

1. 19   **3.** $45
2. 4   **4.** 48

Page 54: Averaging Sales

1. a) $31,000   **b)** $6,200
2. a) $76,000   **b)** $15,200
3. a) $36,000   **b)** $7,200
4. a) $90,000   **b)** $18,000
5.

June Book Sales		
Week	Total Sales	Average Sales
1	$31,000	$6,200
2	$76,000	$15,200
3	$36,000	$7,200
4	$90,000	$18,000

Page 55: Graph the Scales

1.

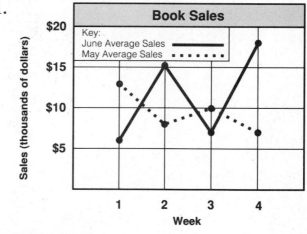

2. Refer to graph.
3. a) May   **c)** May
    b) June   **d)** June
4. June

Page 56: Rank Order

1. 68, 72, 75, 82, 90, 95, 100
2. 850; 859; 1,127; 1,136; 1,454
3. 104,760; 172,540; 232,400; 367,080; 632,000
4. 742; 1,906; 4,252; 10,222; 26,963
5. 75°, 78°, 79°, 86°, 90°, 92°

Page 57: Median and Mode

	Median	Mode
1.	6	6
2.	15	19
3.	138	None
4.	35	17
5.	8	7

Page 58: Find the Median

1. 60   **4.** 136
2. 17   **5.** 77
3. 20

Page 59: Plot the Median

1. a) 92   **d)** 77
b) 85   **e)** 98
c) 71

2.

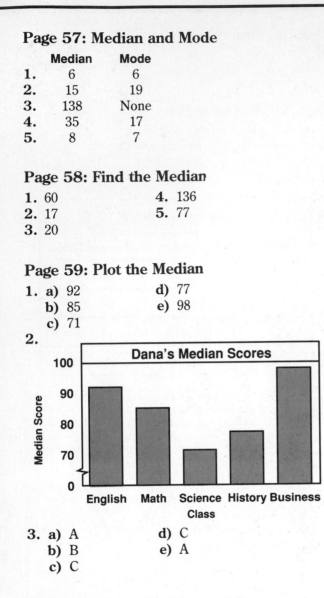

3. a) A   **d)** C
b) B   **e)** A
c) C

Page 60: Mean, Median, and Range

1. a) 78°   **d)** 62° and 73°
b) 81°   **e)** 81°, 86°, and 88°
c) 26°

2. a) 78   **d)** 65, 68, and 73
b) 79   **e)** 85, 87, and 90
c) 25

3. a) $30   **c)** $75
b) $20   **d)** Mean

4. a) 147   **c)** 67
b) 146   **d)** 3

Page 61: Using Data

1. 36   **5. a)** 10
2. 96   **b)** 13
3. a) 6   **6. a)** 5
b) 16   **b)** None
4. a) 5   **7.** Refer to graph. This
b) 16   may also be a bar graph.

Page 61: Using Data (continued)

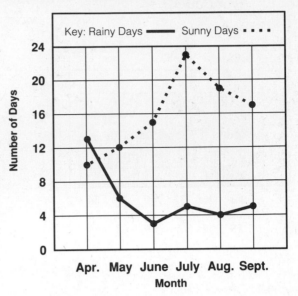

8. Answers will vary.

Page 62: Finding Data

	a)	**b)**	**c)**
1.	117	143	85
2.	149	164	125
3.	117	134	105
4.	144	152	137
5.	176	205	159
6.	131	141	118
7.	199	210	183

Page 63: Graph the Data

1.

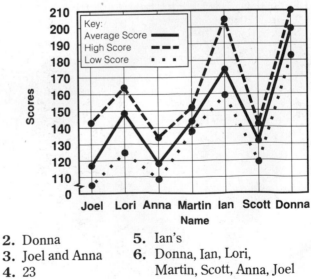

2. Donna   **5.** Ian's
3. Joel and Anna   **6.** Donna, Ian, Lori,
4. 23   Martin, Scott, Anna, Joel

Page 64: Comparing Graphs

1. Health and travel
2. 12%

Page 64: Comparing Graphs (continued)
3. Line graph
4. a) Housing and food
   b) 40%
5. About $72

Page 65: Practice Your Skills
1. a) 75
   b) The score of 75 happened 10 times.
2.

Score	60	65	70	75	80	85	90	95	100
Frequency	4	3	6	10	12	8	6	6	2

3. 57
4. 100
5. 40
6. Answers will vary.

Page 66: Graphs Review
1. Family Budget
2. Food
3. Housing
4. $9,834
5. Education and Total Earnings
6. Yearly earnings in thousands of dollars
7. Education level
8. a) About $14,000
   b) About $22,000
   c) About $32,000
   d) About $42,000
9. a) Game 6
   b) Game 1
10. Answers will vary.
11. $500
12. a) $3,000
    b) $1,250
    c) $2,000
13. $250
14. $6,250